Edwin Zondervan
Process Optimization

Also of Interest

Industrial Process Plants.
Global Optimization of Utility Systems
Ravi Nath, 2024
ISBN 978-3-11-101531-6
e-ISBN 978-3-11-102067-9

Product-Driven Process Design.
From Molecule to Enterprise
2nd Edition
Edwin Zondervan, Cristhian Almeida-Rivera and Kyle Vincent Camarda
(Eds.), 2023
ISBN 978-3-11-101490-6
e-ISBN 978-3-11-101495-1

Sustainable Process Integration and Intensification.
Saving Energy, Water and Resources
3rd Edition
Jiří Jaromír Klemeš, Petar Sabev Varbanov, Sharifah Rafidah Wan Alwi,
Zainuddin Abdul Manan, Yee Van Fan and Hon Huin Chin, 2023
ISBN 978-3-11-078283-7
e-ISBN 978-3-11-078298-1

Multi-level Mixed-Integer Optimization.
Parametric Programming Approach
Styliani Avraamidou and Efstratios Pistikopoulos, 2022
ISBN 978-3-11-076030-9
e-ISBN 978-3-11-076031-6

Edwin Zondervan

Process Optimization

——

Decision-Making Tools for Chemical Engineers

DE GRUYTER

Author
Prof. Dr. ir. Edwin Zondervan
Faculty of Science and Technology
Twente University
Drienerlolaan 5
7500 AE Enschede
The Netherlands
e.zondervan@utwente.nl

ISBN 978-3-11-134208-5
e-ISBN (PDF) 978-3-11-134228-3
e-ISBN (EPUB) 978-3-11-134275-7

Library of Congress Control Number: 2025946414

Bibliographic information published by the Deutsche Nationalbibliothek
The Deutsche Nationalbibliothek lists this publication in the Deutsche Nationalbibliografie;
detailed bibliographic data are available on the internet at http://dnb.dnb.de.

© 2026 Walter de Gruyter GmbH, Berlin/Boston, Genthiner Straße 13, 10785 Berlin
Cover image: wildpixel/iStock/Getty Images Plus
Typesetting: Integra Software Services Pvt. Ltd.

www.degruyterbrill.com
Questions about General Product Safety Regulation:
productsafety@degruyterbrill.com

לשמואל גליקמן

And to
all students,
especially the ones
who join my Tuesday Rugby sessions

Foreword

When I began my academic journey in 2008, I had the extraordinary opportunity to visit some of the world's leading institutions: Universitat Politècnica de Catalunya, Carnegie Mellon University, Denmark Technical University, and Imperial College London, as a visiting scholar. These experiences were transformative, allowing me to assemble a rich toolbox of process optimization techniques that would later become the foundation of my research and teaching. Little did I know then how profoundly these tools would shape my career and my ability to guide the brilliant minds of my PhD students.

In 2012, driven by a desire to share this knowledge in a structured way, I developed an elective Master's course at Eindhoven University of Technology. To my delight, the course resonated with students, and its success encouraged me to refine and expand it over the years, first at the University of Bremen and later, from 2020 onward, at the University of Twente. The enthusiasm of my students and their hunger for practical, applicable knowledge inspired me to take on a new challenge: transforming years of lecture notes, exercises, and insights into a comprehensive textbook.

This book is the culmination of that vision. It has been a labor of love and, at times, a struggle, balancing the demands of academia with the meticulous work of organizing, updating, and expanding the material. But the effort has been worth it. Within these pages, you will find not only the core principles of process optimization but also fresh perspectives on emerging trends, such as the role of artificial intelligence and quantum computing in shaping the future of our field.

I wanted this book to be more than just a technical manual. It is a reflection of my teaching philosophy: rigorous yet accessible, grounded in theory but rich with real-world applications. You'll encounter worked examples, case studies, and even playful anecdotes, in the form of cartoons, that capture the struggles and triumphs of students grappling with optimization problems.

I am grateful to the eminent colleagues* who contributed their expertise through the *"From the Expert"* testimonies. Their insights elevate this book, connecting it to the broader landscape of process systems engineering.

A special thanks goes to Shmuel Glickman, whose unwavering encouragement and belief in this project kept me motivated to see it through to completion.

My hope is that this book brings you both joy and knowledge, that it equips you with decision-making tools to tackle the complex challenges of chemical engineering, and inspires you to push the boundaries of what optimization can achieve. Whether you are a student, a researcher, or a practitioner, I invite you to explore, question, and innovate. . .. *The toolbox is now in your hands.*

Edwin Zondervan
August 2025

https://doi.org/10.1515/9783111342283-202

*PSF Wall of Fame: Top (from left to right): Antonis Kokossis, Ignacio Grossmann, Johan Grievink, Rafiqul Gani, Sigurd Skogestad. Bottom (from left to right): Ana Povoa, Kai Sundmacher, me with my youngest son, Stratos Pistikopoulos, Daniel Lewin.

Contents

1 Introduction to Process Optimization

> The art of optimization lies not in seeking perfection, but in mastering the balance between what is possible and what is practical, where science meets the wisdom of choice.

Process optimization lies at the heart of chemical engineering, serving as a critical tool for improving efficiency, reducing costs, and ensuring *sustainability* in industrial processes. At its core, optimization involves making the best possible decisions within given constraints, whether in designing a reactor, operating a distillation column, or managing an entire production facility. This chapter introduces the fundamental concepts of process optimization, its significance in chemical engineering, and the diverse methodologies employed to solve real-world problems.

1.1 The Role of Optimization in Chemical Engineering

Chemical engineers are tasked with designing, operating, and improving processes that transform raw materials into valuable products. These processes often involve complex interactions between variables, such as temperature, pressure, flow rates, and chemical compositions. Optimization provides a systematic framework for navigating these complexities, enabling engineers to identify the most efficient operating conditions, minimize waste, and maximize profitability.

From the Expert: **Prof. Rafiqul Gani**
PSE for Speed Company, Denmark
Key Laboratory of Smart Manufacturing
in Energy Chemical Processes,
Ministry of Education, East China University of
Science and Technology, Shanghai, China

Solution of Optimization Problems in Process Systems Engineering
1) Personal Insight: Many PSE (Process Systems Engineering)-related problems can be formulated as optimization problems and solved through appropriate numerical solvers. For example, in modeling, the regression of the model parameters; in product or process synthesis, the determination of the optimal network representing the product's molecular structure or process flow-diagram; in process simulation and design, the determination of the optimal values of the design variables; in product design, the determination of the optimal mixture of chemicals representing a desired formulated product; and in control, the determination of the optimal control action based on real-time data. The main issues are understanding the system; representing the system with appropriate models; formulating a consistent optimization problem; and reliably solving the problem with a suitable numerical solver. In the era of AI (artificial intelligence), PSE techniques are augmented by AI techniques to increase the sphere of influence of the PSE

https://doi.org/10.1515/9783111342283-001

technique. For example, ML (machine learning) plays an important role in various forms, such as ensemble modeling, stochastic inference, neural networks, and reinforced learning.

2) Main Achievement in the Field: Three examples of achievements in the field from the PSE area are given here. The first is the development of the process simulator with built-in simulation and/or optimization features. Note that at any time on Earth, somebody somewhere is probably using a process simulator to simulate and/or optimize a process from the chemical and related industrial sectors. The second is the development and availability of efficient numerical solvers (such as NLP, MILP, MINLP, and ML methods), which has allowed the routine solution of various product and/or process synthesis and design-related problems by academia and industry. The third is the use of ML-based methods for accurate data-driven modeling, and the use of RL (reinforced learning)-based optimization, which is finding increasing use in applications of control techniques such as MPC (model predictive control).

3) Challenges and Road Ahead: Three issues and challenges are highlighted here. The first is the selection and application of the optimal solution strategy (workflow and dataflow). That is, a direct solution approach (solve all equations simultaneously, as in the synthesis of heat exchange networks or process synthesis with superstructure-based optimization using simple process models); or a decomposition-based approach (solve subsets of equations in a hierarchical order, as in simultaneous product and process synthesis and design; integrated process synthesis, design, and control). The second is related to the representation of the system with models. That is, the model (size, complexity, form, etc., as in steady-state models for process operation, dynamic models for process control, or distributed parameter models for 3D analysis of equipment design, all within an optimization loop) that affects the solution strategy. The third is related to the selection, implementation, and performance of the numerical solver. Related issues are computer and available computing power; performance of the solver; and limitations related to the need for a good initial estimate. The selection of the numerical solver also affects the solution strategy.

4) A Tip for the Reader: In the era of computers and AI, develop and/or select the augmented PSE technique where the computer addresses the issues of size and speed; AI tackles the complexity through model-based decision-making; numerical methods (including ML methods) provide robust and accurate solutions to the problem; and the human (process engineer) glues everything together and trains the augmented PSE technique to solve a wide range of problems. The issue of automation, as opposed to interactivity with some decisions reserved for built-in AI techniques and others made by humans, could give the best results.

For instance, consider the design of a heat exchanger network. By formulating the problem mathematically, engineers can determine the optimal configuration that minimizes energy consumption while meeting heat transfer requirements. Similarly, in reactor design, optimization techniques help balance reaction kinetics, selectivity, and safety constraints to achieve the desired product yield. Beyond equipment design, optimization plays a pivotal role in supply chain management, production scheduling, and even environmental impact mitigation.

1.2 Key Objectives of Process Optimization

The primary goals of process optimization can be distilled into three overarching themes: *cost reduction, efficiency enhancement*, and *sustainability*. Cost reduction focuses on minimizing capital and operational expenses, such as raw material usage or energy consumption. Efficiency improvement seeks to maximize output per unit of input, whether in terms of product purity, production rate, or resource utilization. Sustainability, an increasingly critical objective, involves optimizing processes to reduce environmental footprints, such as greenhouse gas emissions or water usage, while maintaining economic viability.

These objectives often intersect and sometimes conflict. For example, increasing reactor temperature may improve reaction rates but could also elevate energy costs or safety risks. Such trade-offs necessitate a structured approach to decision-making, where optimization tools provide quantitative insights to balance competing priorities.

1.3 Types of Optimization Problems

Optimization problems in chemical engineering can be broadly classified based on the nature of their variables and constraints. *Linear programming* (*LP*) deals with problems where the objective function and constraints are linear, making it suitable for resource allocation or blending problems. *Nonlinear programming* (*NLP*) addresses more complex scenarios, such as reaction kinetics or thermodynamic equilibria, where relationships between variables are nonlinear.

Integer and *mixed-integer programming* (*MILP/MINLP*) introduces discrete decisions, such as selecting equipment sizes or scheduling batch operations. Multi-objective optimization tackles problems with conflicting goals, like minimizing costs while maximizing efficiency, and employs techniques like Pareto analysis to identify compromise solutions. Finally, optimization under uncertainty accounts for variability in parameters, such as feedstock quality or market demand, using stochastic or robust methods.

1.4 Real-World Applications

The principles of process optimization are applied across diverse industries. In petrochemicals, optimization guides the design of *distillation sequences* to separate hydrocarbon mixtures with minimal energy expenditure. Pharmaceutical companies leverage these tools to optimize batch processes, ensuring compliance with stringent quality standards while reducing production time. In renewable energy, optimization aids in the design of biofuel production pathways that maximize yield and minimize environmental impact.

These applications underscore the versatility of optimization techniques, which adapt to the unique challenges of each sector while adhering to universal mathematical principles. In figure 1.1 student Valerio is starting his optimization class.

Figure 1.1: Student Valerio engaging with the process optimization course.

1.5 Outline of the Book

This book is structured to provide a comprehensive understanding of process optimization, from foundational concepts to advanced applications. Following this introduction, Chapter 2 reviews the mathematical tools essential for optimization, including calculus and linear algebra. Chapter 3 delves into problem formulation, illustrating how to translate real-world scenarios into mathematical models.

Chapters 4 through 6 explore specific optimization methodologies: Linear programming (Chapter 4), nonlinear programming (Chapter 5), and integer programming (Chapter 6). Chapter 7 addresses multi-objective optimization, focusing on trade-offs and Pareto optimality. Chapter 8 examines decision-making under uncertainty, introducing stochastic and robust techniques.

Chapter 9 highlights cutting-edge applications, such as artificial intelligence and quantum computing, while Chapter 10 offers a practical tutorial on using GAMS software for optimization. Each chapter includes worked examples, case studies, and end-of-chapter problems to reinforce learning.

By the end of this book, readers will possess the *theoretical knowledge* and *practical skills* to tackle optimization challenges in chemical engineering, equipping them to make informed, data-driven decisions in their professional endeavors.

1.6 Course Adaptability

This book is designed to cater to both *undergraduate* and *graduate* students, with its content adaptable to different levels of academic rigor. Below is a suggested breakdown of how the material can be structured for introductory and advanced courses.

1.6.1 Undergraduate Course (Introductory Optimization)

For an undergraduate course, the focus should be on building foundational knowledge and practical problem-solving skills. The following chapters are particularly suitable:
–   Chapter 1 provides an overview of optimization's role in chemical engineering.
–   Chapter 2 reviews essential calculus and linear algebra concepts.
–   Chapter 3 teaches students how to translate real-world problems into mathematical models.
–   Chapter 4 covers fundamental LP techniques, including the Simplex method and sensitivity analysis.
–   Chapter 5 introduces gradient-based methods and constrained optimization.
–   Chapter 10 provides hands-on training in advanced optimization software.

Suggested Approach:
–   Emphasize graphical solutions and manual calculations for LP problems.
–   Use case studies (e.g., reactor optimization and heat exchanger design) to illustrate applications.
–   Assign end-of-chapter problems that reinforce basic concepts.
–   Introduce software tools (Excel Solver and basic GAMS) for hands-on learning.

1.6.2 Graduate Course (Advanced Optimization)

For graduate students, the focus shifts toward advanced methodologies, uncertainty analysis, and cutting-edge applications. The following chapters are recommended:
–   Chapter 5 dives deep into Newton's method, Lagrange multipliers, and penalty functions.
–   Chapter 6 covers branch-and-bound methods for discrete optimization.
–   Chapter 7 expands on advanced techniques like ε-constraint and weighted-sum methods.
–   Chapter 8 introduces stochastic programming, robust optimization, and Monte Carlo methods.
–   Chapter 9 explores AI, metaheuristics, and quantum computing in optimization.
–   Chapter 10 provides hands-on training in advanced optimization software.

Suggested Approach:
- Focus on numerical methods and algorithmic implementations.
- Assign research-oriented case studies (e.g., enterprise-wide optimization and sustainable process design).
- Incorporate real-world datasets for stochastic and robust optimization problems.
- Encourage the use of Python-based optimization libraries (SciPy and Pyomo) alongside GAMS.

Further Reading

Biegler, L. T. (2010). *Nonlinear Programming: Concepts, Algorithms, and Applications to Chemical Processes*. SIAM.

Edgar, T. F., Himmelblau, D. M., & Lasdon, L. S. (2001). *Optimization of Chemical Processes* (2nd ed.). McGraw-Hill.

Tawarmalani, M., & Sahinidis, N. V. (2002). *Convexification and Global Optimization in Continuous and Mixed-Integer Nonlinear Programming*. Kluwer Academic Publishers.

Floudas, C. A. (1995). *Nonlinear and Mixed-Integer Optimization: Fundamentals and Applications*. Oxford University Press.

Grossmann, I. E., & Trespalacios, F. (2003). *Systematic modeling of discrete-continuous optimization models through generalized disjunctive programming*. AIChE Journal, 49(7), 1607–1621.

2 Mathematical Foundations for Optimization

> In optimization, every derivative and matrix tells a story of efficiency waiting to be unlocked.

In this chapter, we will review some topics from *calculus*, *linear algebra*, and *numerical analysis*. Those are the components that are needed to understand, solve, and analyze optimization models. We will look into calculus tools, especially derivatives, for optimality analysis. We will also recap on linear algebra, i.e., matrix inversion, rank, determinant, etc., which we need to formulate optimization problems. We will also refresh the main idea of convexity as a diagnostic for determining whether or not an optimization problem has a global optimum. We will finalize this chapter with a compact review of numerical methods to solve equations.

2.1 Review of Calculus for Optimization

2.1.1 Derivatives and Partial Derivatives

The *derivative* of a function $f(x)$ at a point $x = a$ is denoted as $f'(a)$ or $\frac{df}{dx}\big|_a$. It measures the rate of change of f with respect to x. Geometrically, it represents the slope of the line that touches the curve $y = f(x)$ at $x = a$.

The definition of a derivative is given by:

$$f'(a) = \lim_{h \to 0} \frac{f(a+h) - f(a)}{h}$$

For multivariable functions, e.g., $f(x,y)$, *partial derivatives* ($\partial f/\partial x$, $\partial f/\partial y$) measure the rate of change along one variable while holding others constant.

Suppose we want to find the slope of a tangent on $f(x) = x^2$ on any given value of x. We could use the equation above:

$$f'(x) = \lim_{h \to 0} \frac{f(x+h) - f(x)}{h} = \lim_{h \to 0} \frac{(x+h)^2 - x^2}{h} = \cdots$$

$$= \lim_{h \to 0} \frac{(x^2 + 2hx + h^2) - x^2}{h} = \lim_{h \to 0} 2x + h^2 = 2x$$

For any given value of x, we now know that the slope of the tangent on that point equals $2x$. We can do this for any type of function $f(x)$.

Now, let us look at a practical example: let us consider the following Arrhenius reaction rate equation:

$$k(T) = k_0 e^{-\frac{E_A}{RT}}$$

https://doi.org/10.1515/9783111342283-002

For a pre-exponential factor $k_0 = 1 \times 10^8$ (s^{-1}), an activation energy $E_A = (50 \text{ kJ/mol})$ and using the gas constant $R = 8.314 \left(\frac{J}{\text{mol}} \cdot K \right)$ We can plot the reaction rate for different temperatures, in Figure 2.1, the blue line. We are now interested in the tangent to the reaction rate at $T = 350$ K (the black dot in the figure). This is the derivative at this point, showing how sensitively the reaction rate responds to small temperature changes. The tangent is represented by the red line.

We can approximate the slope of the tangent at $T = 350$ K with the following definition:

$$k'(350) = \lim_{h \to 0} \frac{k(350 + h) - k(350)}{h} \approx 1.75 \times 10^4 \ (\text{s}^{-1} \ \text{K}^{-1})$$

Derivative as Tangent Slope: Reaction Rate Sensitivity
Arrhenius Kinetics Example

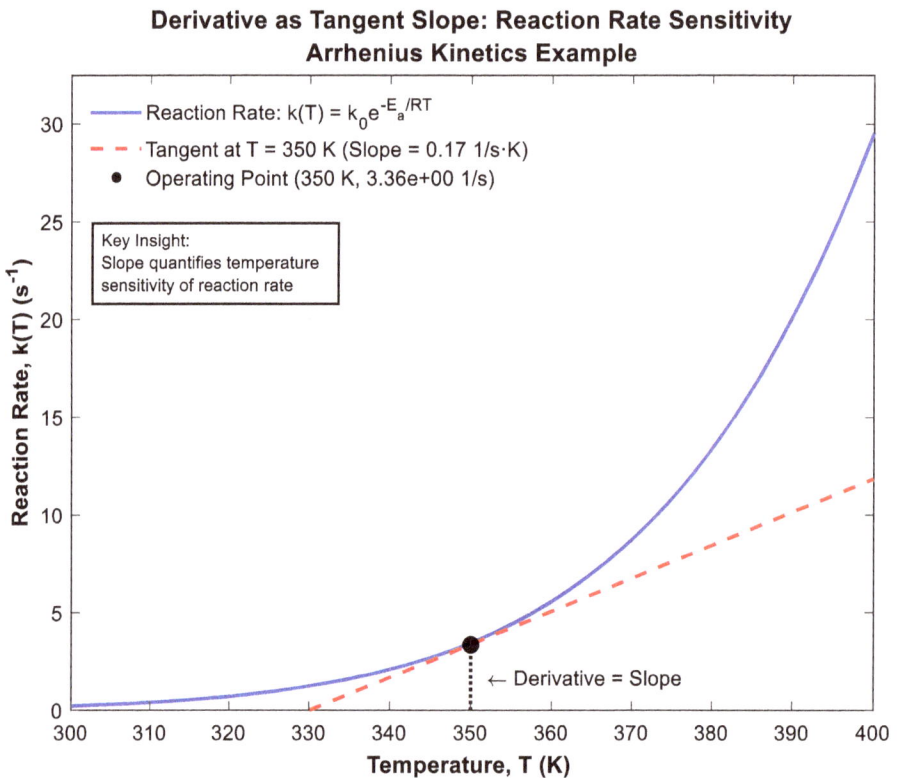

Figure 2.1: The derivative (slope of the red tangent line) quantifies how sensitively the reaction rate responds to temperature changes at $T = 350$ K.

Derivatives *quantify* sensitivity: how does a system's output respond to a tiny change in the input? We use derivatives everywhere in chemical engineering: mass and energy balances are describing accumulation over time (as derivatives); we use derivatives to see how yield might change if we have temperature shifts (reaction engineering), or how change in volume affects pressure in a compressor (to get pressure–volume trade-offs for process control) or to see how profit is affected by changes in production rates (process economics).

We can calculate derivatives in an analytical fashion by using calculus rules (e.g., the *chain rule* or *quotient rule*), as shown for our parabola example. We would have found for our reaction rate example that the derivative for any T can be found from:

$$\frac{dk}{dT} = k(T)\frac{E_A}{RT^2}$$

Or we could use numerical approximations, using *finite differences*, where we apply the definition of a limit for a very small value of h.

When computing derivatives, please 1) check for unit consistency (derivatives must have physically meaningful units), 2) be alert for discontinuities (non-smooth functions, such as, for example, phase transitions, require piecewise derivatives), and 3) use directional derivatives in high-dimensional systems to explore critical paths.

We need to evaluate derivatives in optimization, because it turns out that at an *extremum* (a *minimum* or *maximum*) the derivative of a function turns out to be zero.

Here is a small example (probably you have seen it in high school). Suppose a manufacturer of soup wants to minimize the material costs for soup tin cans. He assumes that the costs are proportional to the amount of tin he is using for the can, and he wants to make sure that a can can contain a liter of soup.

We could now calculate the costs by determining the surface of tin that we need to create a cylinder with a radius r and a height h:

$$C = 2\pi r^2 + 2\pi rh$$

We can also calculate the volume of this can, which has to be one liter:

$$V = \pi r^2 h = 1$$

We can rewrite this equation in terms of h :

$$h = \frac{1}{\pi r^2}$$

We could substitute this into our cost equation:

$$C = 2\pi r^2 + \frac{2}{r}$$

We now have an expression that gives the costs as a function of the radius of the can. Figure 2.2 shows this relationship.

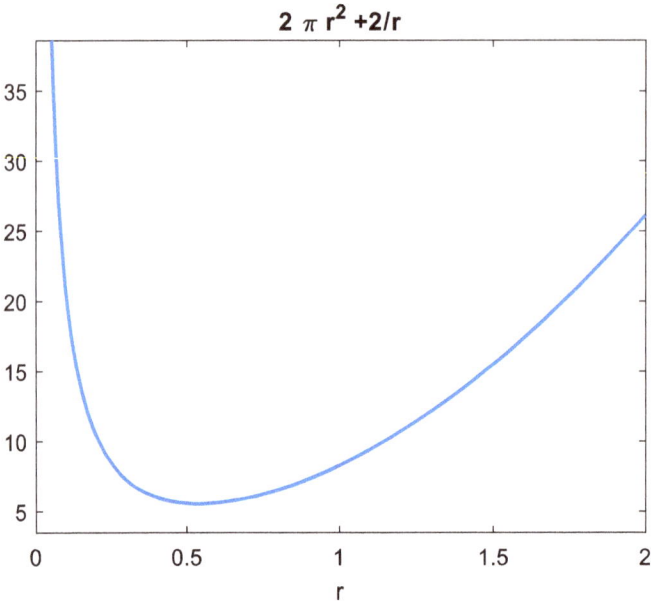

$$2 \pi r^2 +2/r$$

Figure 2.2: Costs of a can as a function of its radius.

As Figure 2.2 shows, the costs are lowest around a radius of more or less 0.5 (dm). At this point, the slope of the tangent has to be equal to zero (a horizontal line). We can analytically derive the derivative of the cost function with respect to the radius:

$$\frac{dC}{dr} = 4\pi r - \frac{2}{r^2}$$

We can plot the derivative of the costs to the radius in Figure 2.3. And now we have to determine for which value of r the derivative becomes zero:

$$\frac{dC}{dr} = 0$$

We can find r easily:

$$4\pi r = \frac{2}{r^2} \Leftrightarrow 2\pi = r^{-3} \Leftrightarrow \quad r = \left(\frac{1}{2\pi}\right)^{\frac{1}{3}} \Leftrightarrow r = 0.54 \ (dm)$$

We can now also compute the height of the can:

$$h = \frac{1}{\pi r^2} = 1.08 \ (dm)$$

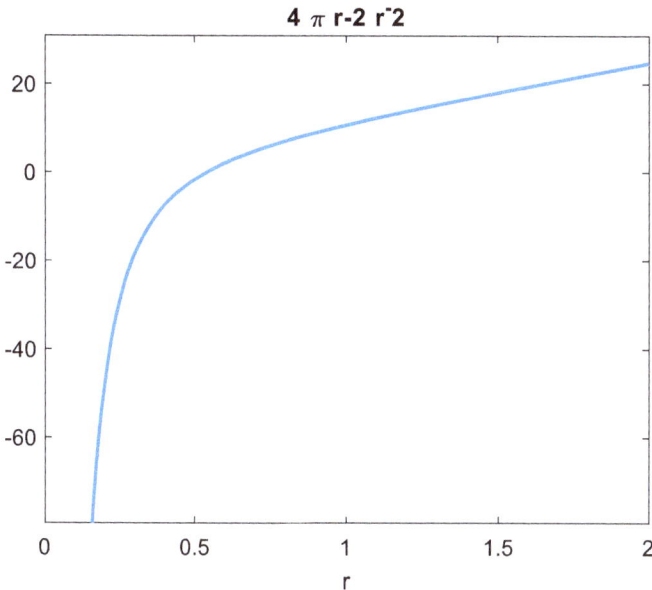

$4\,\pi\,\text{r-2 r}^{\displaystyle\tilde{}}2$

Figure 2.3: The slope of the costs for different radii.

Often, a function might depend on multiple variables, and we might want to evaluate how the derivative behaves with respect to these variables. In this case, we speak of partial derivatives.

As an example, suppose we have a chemical mixer that produces an output reaction rate R (mol/s) that depends on the temperature T(K) and the concentration C(mol/L) given by an empirical model:

$$R(T,C) = 2.5T^{0.5}C^{1.2}$$

We can now compute the partial derivatives to analyze how R responds to changes in T and C independently.

To compute the sensitivity to temperature, we calculate the partial derivative with respect to temperature (for a given concentration):

$$\frac{\partial R}{\partial T} = 2.5 \cdot 0.5 \cdot T^{-0.5} \cdot C^{1.2}$$

And the sensitivity to concentration is (for a given temperature is):

$$\frac{\partial R}{\partial C} = 2.5 \cdot 1.2 \cdot T^{0.5} \cdot C^{0.2}$$

We can now use the partial derivatives for a sensitivity analysis. Suppose we are operating at $T = 400$ K and $C = 2.0$ mol/L. We can now use the partial derivatives to see the effect of temperature and concentration on the reaction rate. It turns out that a 1 K increase in T raises the reaction rate by 0.091 mol/s, while a 1 mol/L increase in C raises R by 60.8 mol/s.

This is rather useful information. For process control, it suggests that concentration adjustments are more effective for rapid output changes. For safety considerations, it is noted that runaway reactions can occur if C is not tightly controlled. For process optimization, we can use the gradients $\nabla R = [\partial R/\partial T, \ \partial R/\partial C]$ in algorithms like gradient ascent to maximize R.

2.1.2 Gradients and Directional Derivatives

As shown in the previous section, a function might depend on multiple variables. This leads to the concept of partial derivatives, or the gradient. The *gradient* of a scalar function $f(x)$ where $x = [x_1, x_2, \ldots, x_n]^T$ is a vector of its partial derivatives:

$$\nabla f = \left[\frac{\partial f}{\partial x_1}, \frac{\partial f}{\partial x_2}, \ldots, \frac{\partial f}{\partial x_n} \right]^T$$

The gradient points in the direction of the *steepest ascent* of f. Its magnitude $||\nabla f||$ quantifies the rate of increase.

In addition, we might speak of *directional derivatives*. A directional derivative measures the rate of change of a function f along a unit vector u:

$$D_u f = \nabla f \cdot u = ||\nabla f|| \cdot \cos \theta$$

where θ is the angle between ∇f and u. The directional derivative is maximized when u aligns with ∇f, in other words when $\cos \theta = 1$.

Example: consider a reactor with yield $Y(T, P) = 2T^{0.6}P^{0.4}$, where T is the temperature in K and P is the pressure in (atm). We can compute the gradient:

$$\nabla Y = \left[\frac{\partial Y}{\partial T}, \frac{\partial Y}{\partial P} \right] = \left[1.2T^{-0.4}P^{0.4}, 0.8T^{0.6}P^{-0.6} \right]$$

At $T = 500$ K and $P = 10$ atm:

$$\nabla Y = [0.072, 0.101]$$

So, for the *steepest ascent*, to maximize the yield increase T and P in a [0.072"0.101] ratio $\approx 1{:}1.4$. Or, regarding the directional derivative: Along $u = [1, 0]$ (vary only T), $D_u Y = 0.072$.

So far, we have seen the partial derivatives of one function f with respect to multiple variables $x_1, x_2, \ldots, x_n$ We could also study the derivatives of multiple functions F that depend on multiple variables x:

$$F(x) = \begin{bmatrix} F_1(x_1, \ldots, x_n) \\ \vdots \\ F_m(x_1, \ldots, x_n) \end{bmatrix}$$

The *Jacobian* matrix J is an $m \times n$ matrix where each entry J_{ij} is the partial derivative of F_i with respect to x_j (where $i = 1, \ldots, m$ and $j = 1, \ldots, n$):

$$J = \begin{bmatrix} \frac{\partial F_1}{\partial x_1} & \cdots & \frac{\partial F_1}{\partial x_n} \\ \vdots & \ddots & \vdots \\ \frac{\partial F_m}{\partial x_1} & \cdots & \frac{\partial F_m}{\partial x_n} \end{bmatrix}$$

Figure 2.4: Bilal, the matrix whisperer: how to deal with wobbling Jacobians.

Figure 2.4 shows grad student Bilal who struggles with Newton's method and Jacobians Here is a small example. Let us consider a reactor in which two reactions occur:

$$A \rightarrow B(r_1 = k_1 C_A)$$

$$B \rightarrow C(r_2 = k_2 C_B)$$

The governing *ODE*s (mass balances) are:

$$F_1 = \frac{dC_A}{dt} = -k_1 C_A$$

$$F_2 = \frac{dC_B}{dt} = k_1 C_A - k_2 C_B$$

$$F_3 = \frac{dC_C}{dt} = k_2 C_B$$

The Jacobian of this system is:

$$J = \begin{bmatrix} \frac{\partial F_1}{\partial C_A} & \frac{\partial F_1}{\partial C_B} & \frac{\partial F_1}{\partial C_C} \\ \frac{\partial F_2}{\partial C_A} & \frac{\partial F_2}{\partial C_B} & \frac{\partial F_2}{\partial C_C} \\ \frac{\partial F_3}{\partial C_A} & \frac{\partial F_3}{\partial C_B} & \frac{\partial F_3}{\partial C_C} \end{bmatrix} = \begin{bmatrix} -k_1 & 0 & 0 \\ k_1 & -k_2 & 0 \\ 0 & k_2 & 0 \end{bmatrix}$$

The third column contains only zeros, as C does not appear in F_1, F_2, or F_3. In other words, C does not influence A or B. We can use the Jacobian to determine whether the reactor reaches steady state (by examining whether the eigenvalues are negative, see Sections 2.1.3 and 2.2.3), we could use *Newton's method* to find the steady state concentrations by solving $F = 0$, and we can use the Jacobian in a sensitivity analysis to see how perturbations in the concentrations propagate through the system.

2.1.3 Hessian Matrix and Curvature

The *Hessian* matrix effectively contains the second-order derivatives of functions (as the Jacobian contains the first-order derivatives). As you recall, the first derivative can be used to determine an extremum of a function. The second derivative can be used to obtain information on the curvature of a function.

For a function $f(x)$, the Hessian is an $n \times n$ matrix:

$$H(x) = \begin{bmatrix} \frac{\partial^2 f}{\partial x_1^2} & \frac{\partial^2 f}{\partial x_1 \partial x_2} & \cdots \\ \frac{\partial^2 f}{\partial x_2 \partial x_1} & \frac{\partial^2 f}{\partial x_2^2} & \cdots \\ \vdots & \vdots & \ddots \end{bmatrix}$$

If f is twice continuously differentiable, then:

$$\frac{\partial^2 f}{\partial x_i \partial x_j} = \frac{\partial^2 f}{\partial x_j \partial x_i}$$

This is the so-called symmetry property.

If we look at a 1D example, $f''(x) > 0$ implies *convexity*. The Hessian holds a similar concept for multidimensional, via eigenvalues.

If we look at the extremum x^* (where $\nabla f(x^*) = 0$), the *eigenvalues* tell us whether we are dealing with a minimum, or maximum, or a saddle point, see Table 2.1.

Table 2.1: Hessian and critical point.

Eigenvalues of Hessian	Critical point type	Curvature analogy
All $\lambda_i > 0$	Strict local minimum	Bowl-shaped
All $\lambda_i < 0$	Strict local maximum	Inverted bowl
Mixed signs	Saddle point	Mountain pass
Some $\lambda_i = 0$	Degenerate	Flat or higher-order

Here is an example: Let us analyze the critical points of $f(x,y) = x^3 + y^3 - 3xy$

We can find the critical points via: $\nabla f = \begin{bmatrix} 3x^2 - 3y, & 3y^2 - 3x \end{bmatrix} = 0 \rightarrow (0,0)$ and $(1,1)$.

We calculate the Hessian:

$$H = \begin{bmatrix} 6x & -3 \\ -3 & 6y \end{bmatrix}$$

We evaluate the Hessian at $(1,1)$:

$$H(1,1) = \begin{bmatrix} 6 & -3 \\ -3 & 6 \end{bmatrix}$$

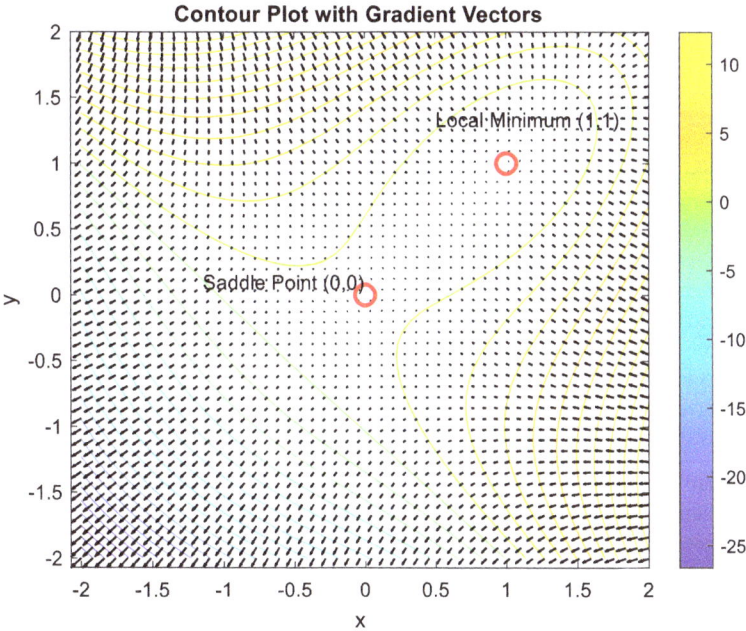

Figure 2.5: Contour plot of f as a function of x and y showing the saddle point and local minimum.

The eigenvalues of this matrix are: $\lambda_1 = 3$ and $\lambda_2 = 9$, which are both positive, meaning this is a local minimum.

At $(0, 0)$ the eigenvalues of the Hessian are $\lambda_1 = 3$ and $\lambda_2 = -3$ which indicate a saddle point. Figure 2.5 shows a contour plot of as function of x and y.

Mugshot Prof. Kai Sundmacher
From the Expert: **Prof. Kai Sundmacher**
Max Planck Institute for Dynamics of Complex Technical Systems
Otto Von Guericke University of Magdeburg, Germany

Dynamic Optimization for Process Intensification

1) Personal Hook/Insight: As a student, I was introduced to optimization methods by my future doctoral supervisor Ulrich Hoffmann, now Emeritus Professor of Chemical Engineering at Clausthal University of Technology. Among other things, he explained to me the features of linear and non-linear programming, how to deal with equality and inequality constraints, the basic idea behind George Dantzig's simplex method, and that 'optimization' simply means 'decision making'. This implies that if you want to optimize a chemical process, you first have to define suitable decision variables and an objective function. If this is done clumsily, you often end up with complex, highly non-linear optimization problems. However, if you do it skillfully, e.g., by using certain groups of variables for decision-making instead of using directly the process variables, you often end up with tasks that can be solved numerically in a relatively short time.

2) Main Achievement in the Field: There is now a comprehensive toolbox available for process intensification (PI). In the period 2000–2010, experts endeavored to systematize the various PI methods and organize their use according to rational criteria, under the motto "PI: from art to science." My group's contribution to these efforts was the introduction of the elementary process functions (EPF) methodology, which can be used to quantitatively assess the potential of process intensification measures. The first step in EPF is to determine the optimal trajectory of a Lagrangian matter element in the thermodynamic state space, whereby the element is controlled by the material and energy exchange fluxes with the environment. Mathematically, this leads to an optimal control problem (OCP), i.e., an optimization problem with path constraints in the form of differential equations. The solution to the OCP provides the optimal trajectory of the matter element. Subsequently, we investigate which controls have the strongest influence on this trajectory and the objective function (e.g., selectivity and productivity). This can then be used to translate the optimal trajectory into a practical design concept for a process unit. In this way, e.g., continuously operated chemical reactors can be developed in which the dosing of the reactants and the heating/cooling is controlled along the reactor coordinate in such a way that undesirable side reactions are strongly suppressed. This can drastically improve the raw material efficiency of a chemical production process.

3) Challenges and Road Ahead: The performance of chemical processes can be improved not only the operational variables, but also by structural variables. Important examples are auxiliary molecular agents used in reaction processes (catalysts, solvents, additives, etc.), auxiliary materials used in separation processes (membranes, adsorbents, etc.), and the interconnection of process units to form the overall production process. The simultaneous optimization of auxiliary materials and process materials together with the operating conditions of the processes has been pursued in academic research (CAMPD: Computer-aided Molecular, Material and Process Design). However, due to the complexity of the CAMPD problems to be solved, this has not yet found its way into industrial practice. Advancing this remains a task for the coming years. Another future task will be to embed data-driven and hybrid submodels instead of purely mechanistic models for the thermodynamic properties of species and the kinetics of chemical reactions, mass and heat transport (e.g., in the form of ANNs) into the OCP problems mentioned above.

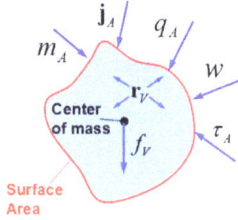

States: $\mathbf{x}^T(t) = [T(t), p(t), \mathbf{y}(t), \mathbf{v}(t)]$

Fluxes: $\mathbf{j}^T = [m_A, \mathbf{j}_A, \mathbf{r}_V, \tau_A, f_V, q_A, w]$

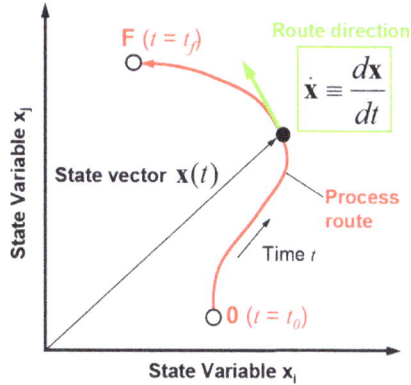

- Consider a matter element travelling in the thermodynamic state space.
- Control its pathway by the action of mass, energy & momentum fluxes.
- Find the optimal trajectory between defined start and end points.
- Translate this trajectory to a technically realizable process design.

4) A Tip for the Reader: Once you understand the mathematical properties and numerical solution of LPs, NLPs, and MINLPs, you have nothing to fear from OCPs. This is because simultaneous parameterization of the control variables and state variables allows an OCP to be converted into an NLP. And if you then want to extend is to CAMPD, you end up with a MINLP. This is how everything is ultimately linked together.

2.2 Linear Algebra Essentials

We need to review a couple of linear algebra essentials, as it turns out that process optimization problems are often represented using matrix/vector notation.

2.2.1 Matrices and Vectors

A *vector* is an ordered list of numbers (for example, concentrations, flow rates), The notation is: $x = [x_1, x_2, \ldots, x_n]^T$. A matrix is a rectangular array that describes linear transformations (think of stoichiometric coefficients, Jacobians). The notation for a matrix is:

$$A = \begin{bmatrix} a_{11} & \cdots & a_{1j} \\ \vdots & \ddots & \vdots \\ a_{i1} & \cdots & a_{ij} \end{bmatrix}$$

The key operations in linear algebra are the multiplications between matrices and vectors:

$$Ax = \begin{bmatrix} a_{11}x_1 + \cdots + a_{1n}x_n \\ \vdots \\ a_{m1}x_1 + \cdots + a_{mn}x_n \end{bmatrix}$$

We can use the so-called *Euclidean*, or *2-norm* for error analysis:

$$\|x\|_2 = \sqrt{x_1^2 + \cdots + x_n^2}$$

And the *Frobenius* norm for matrix sensitivity:

$$\|A\|_F = \sqrt{\sum_i^j a_{ij}^2}$$

There are a number of special matrices in process engineering, for example, the *diagonal matrix*, where $A_{ij} = 0$ $(i{\neq}j)$ (e.g., when describing heat capacity in energy balance systems), the *symmetric matrix* $(A = A^T)$ (e.g., the Hessian matrix discussed in Section 2.1.3) and the *sparse matrix*, where most entries are zero (this, for instance, occurs in process flowsheets with limited connectivity).

2.2.2 Linear Systems and Inverses

To solve a system of *linear equations* $(Ax = b)$, we could use the *inverse matrix*. The inverse of a matrix A is defined such that if we multiply A by its inverse, it should give us the *identity matrix* (a diagonal matrix with only ones in the diagonal):

$$AA^{-1} = I, \ A^{-1}A = I$$

If we use this definition of the inverse, we can solve a linear system:

$$Ax = b \rightarrow x = A^{-1}b$$

But how to compute an inverse? We can calculate the inverse of an $n \times n$ by using *cofactors*:

$$A = \frac{1}{\det(A)} C^T$$

Suppose we want to calculate the co-factors of the following matrix:

$$A = \begin{bmatrix} 1 & 1 & 1 \\ 2 & 1 & 3 \\ 3 & 1 & 6 \end{bmatrix}$$

A co-factor matrix has the same size as the original matrix A. A co-factor is defined to be the *determinant* of the stuff left over when you cover up the row and column of the element in question. We then multiply by ±1 alternating with the position of the element.

Suppose we want to calculate C_{11}, we cover up the row and column of the element in question:

$$\begin{bmatrix} 1 & \ast & \ast \\ \ast & 1 & 3 \\ \ast & 1 & 6 \end{bmatrix}$$

And compute the determinant of the stuff that is left over:

$$C_{11} = + \det \begin{pmatrix} 1 & 3 \\ 1 & 6 \end{pmatrix} = + (6 \times 1 - 3 \times 1) = 3$$

Let us compute C_{12}:. Cover up the row and column of the element in question:

$$\begin{bmatrix} \ast & 1 & \ast \\ 2 & \ast & 3 \\ 3 & \ast & 6 \end{bmatrix}$$

And compute the determinant of the stuff that is left over:

$$C_{12} = + \det \begin{pmatrix} 2 & 3 \\ 3 & 6 \end{pmatrix} = - (6 \times 2 - 3 \times 3) = -3$$

We can repeat this procedure to compute the complete co-factor matrix:

$$C = \begin{bmatrix} 3 & -3 & -1 \\ -5 & 3 & 2 \\ -1 & 2 & -1 \end{bmatrix}$$

To find the inverse, we should also calculate the determinant of matrix A. The determinant is rather important. If the determinant is zero, it means the inverse of the matrix does not exist (we speak of a *singular* matrix).

We have just seen how the determinant of a 2×2 can be calculated. For larger matrices, we do a similar calculation procedure to calculating co-factors. We select any row or column and multiply it by its co-factor. Suppose we use the first row:

$$\det \begin{bmatrix} 1 & 1 & 1 \\ 2 & 1 & 3 \\ 3 & 1 & 6 \end{bmatrix} = +1 \det \begin{pmatrix} 1 & 3 \\ 1 & 6 \end{pmatrix} - 1 \det \begin{pmatrix} 2 & 3 \\ 3 & 6 \end{pmatrix} + 1 \det \begin{pmatrix} 2 & 1 \\ 3 & 1 \end{pmatrix} = -1$$

We could use also use the third column instead as a basis, which will give the same result

$$\det \begin{bmatrix} 1 & 1 & 1 \\ 2 & 1 & 3 \\ 3 & 1 & 6 \end{bmatrix} = +1 \det(1) - 3 \det \begin{pmatrix} 1 & 1 \\ 3 & 1 \end{pmatrix} + 6 \det \begin{pmatrix} 1 & 1 \\ 2 & 1 \end{pmatrix} = -1$$

Suppose we want to solve:

$$\begin{bmatrix} 1 & 1 & 1 \\ 2 & 1 & 3 \\ 3 & 1 & 6 \end{bmatrix} \begin{bmatrix} x_1 \\ x_2 \\ x_3 \end{bmatrix} = \begin{bmatrix} 4 \\ 7 \\ 5 \end{bmatrix}$$

We can now determine the inverse via:

$$A^{-1} = \frac{1}{\det(A)} C^T$$

And then substitute in:

$$x = A^{-1} b$$

which will give:

$$x = \frac{1}{\det(A)} C^T b = \frac{1}{-1} \begin{bmatrix} 3 & -3 & -1 \\ -5 & 3 & 2 \\ -1 & 2 & -1 \end{bmatrix}^T \begin{bmatrix} 4 \\ 7 \\ 5 \end{bmatrix} = \begin{bmatrix} 13 \\ -4 \\ -5 \end{bmatrix}$$

The solution of a system of linear equations may or may not exist, and it may or may not be unique. The existence of solutions can be determined by comparing the *rank* of the matrix A with the rank of the augmented matrix A|b. The rank of a matrix is the number of linearly independent columns, i.e., columns that cannot be expressed as a linear combination of the other columns of a matrix.

The calculation of an inverse with co-factors and determinants is, of course, a computationally very expensive procedure. There are numerical procedures to determine the solutions to linear systems iteratively. *Gaussian elimination* is a procedure where matrix A is treated with row operations in such a way that it turns into a triangular matrix. Such system can be solved directly with substitutions. The Jacobi method updates the solution vector x in a structural way (which is in terms of memory use more effective as the Gaussian elimination method).

However, since the rise of computers, there have been superior algorithms for matrix inversions: to name *LU factorization*, which works with dense, one-off solve type of problems, *QR factorization*, which works with rectangular, and rank-deficient systems or *Cholesky factorization*, which is very effective for symmetric positive definite matrix systems.

In *MATLAB* you could use the *backslash* operator, which uses LU/QR automatically: x = A\b. In Python you could turn to x = np.linalg.solve(A,b).

Here is a small example for a MATLAB implementation:

```
A = [2 -1; -1 2];   % Symmetric positive definite
b = [1; 0];
x = A\b;            % LU/Cholesky auto-selected
cond_A = cond(A);   % Condition number (~3 for this A)
```

And a Python implementation:

```
import numpy as np
from scipy.linalg import solve
A = np.array([[2, -1], [-1, 2]])
b = np.array([1, 0])
x = solve(A, b)  # Uses LU under the hood
cond_A = np.linalg.cond(A)
```

2.2.3 Eigenvalues and Positive Definiteness

Matrices can be characterized by their *eigenvectors e* and eigenvalues λ. For a matrix *A* holds that:

$$Ae = \lambda e \Leftrightarrow$$

$$Ae - \lambda e = 0 \Leftrightarrow$$

$$Ae - I\lambda e = 0 \Leftrightarrow$$

$$(A - I\lambda)e = 0$$

This means that

$$\det(A - I\lambda) = 0$$

Let us consider an example. We want to determine the eigenvalues of the following matrix:

$$A = \begin{bmatrix} 1 & 0 & 1 \\ 0 & 1 & 0 \\ 1 & 0 & 1 \end{bmatrix}$$

We can now see that:

$$\det(A - I\lambda) = \det\left(\begin{bmatrix} 1 & 0 & 1 \\ 0 & 1 & 0 \\ 1 & 0 & 1 \end{bmatrix} - \begin{bmatrix} \lambda & 0 & 0 \\ 0 & \lambda & 0 \\ 0 & 0 & \lambda \end{bmatrix} \right) = \det\left\{ \begin{matrix} 1-\lambda & 0 & 1 \\ 0 & 1-\lambda & 0 \\ 1 & 0 & 1-\lambda \end{matrix} \right\} = 0$$

We know how to calculate the determinant of a 3×3 matrix:

$$\det(A - I\lambda) = (1-\lambda)\det\begin{bmatrix} 1-\lambda & 0 \\ 0 & 1-\lambda \end{bmatrix} - 0\det\begin{bmatrix} 0 & 0 \\ 1 & 1-\lambda \end{bmatrix} + 1\det\begin{bmatrix} 0 & 1-\lambda \\ 1 & 0 \end{bmatrix}$$

$$= (1-\lambda)\left[(1-\lambda)^2 - 0\right] + [0 - 1(1-\lambda)] = 0$$

This is a third-order polynomial; we need to find its roots: $\lambda_1 = 0$, $\lambda_2 = 1$ and $\lambda_3 = 2$

The eigenvalues will tell us something about system stability (when we are solving ODE's), and in optimization the eigenvalues tell us whether we are dealing with (local) maxima, minima, or saddle points (remember the Hessian?).

2.3 Convexity and Concavity

2.3.1 Convex Sets and Functions

Convexity is an important concept in optimization. Convexity guarantees the mathematical nature of solutions. A set $C \in \mathbb{R}^n$ is called convex if for any two $x, y \in C$ points the line segment connecting them lies entirely within C. Formally, this means:

$$\theta x + (1 - \theta)y \in C \; \forall \; \theta \in [0, 1]$$

In chemical engineering, the feasible operating region is defined by linear constraints (e.g., mass balances $Ax = b$ and capacity limits $x \leq x^U$) which always forms a convex set. This property is critical because it ensures that any local optimum within such region is also the global optimum. Figure 2.6 shows different examples of convex and non-convex sets.

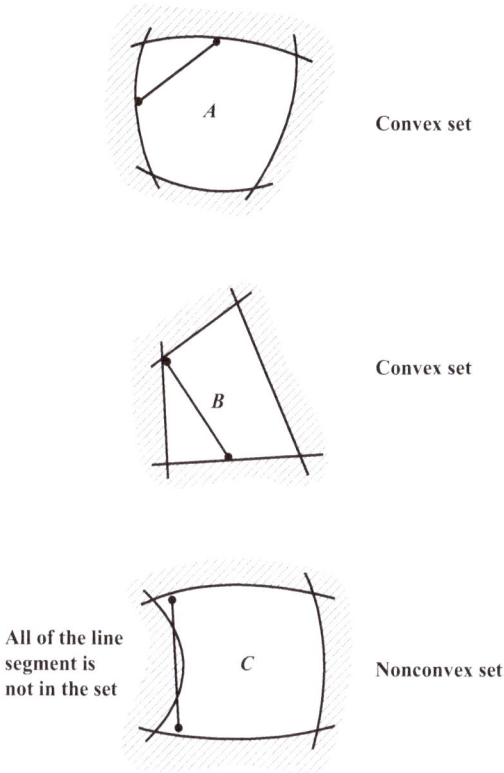

Convex set

Convex set

All of the line
segment is
not in the set

Nonconvex set

Figure 2.6: Convex and non-convex sets.

A function $f: C \to \mathbb{R}$ is convex, if its domain C is convex and *Jensen's inequality* holds:

$$f\left(\theta x_{(1-\theta)y}\right) \leq \theta f(x) + (1-\theta)f(y), \quad \forall x, y \in C, \theta \in [0,1]$$

Geometrically, this implies the function's graph lies below its cord. Figure 2.6 shows an example of a convex and a non-convex function.

For differentiable functions, an equivalent *first-order condition states* that f is convex if and only if:

$$f(y) \geq f(x) + \nabla f(x)^T (y - x) \quad \forall x, y \in f$$

This inequality means the first-order Taylor approximation is a global under-estimator of f. In process systems, convex functions include idealized reactor yield models (quadratic in conversion) and linear pressure-drop relationships, while non-convex examples arise in phase equilibrium calculations or reaction kinetics with inhibition times. In figure 2.7 is an example given of a convex- and a non-convex function.

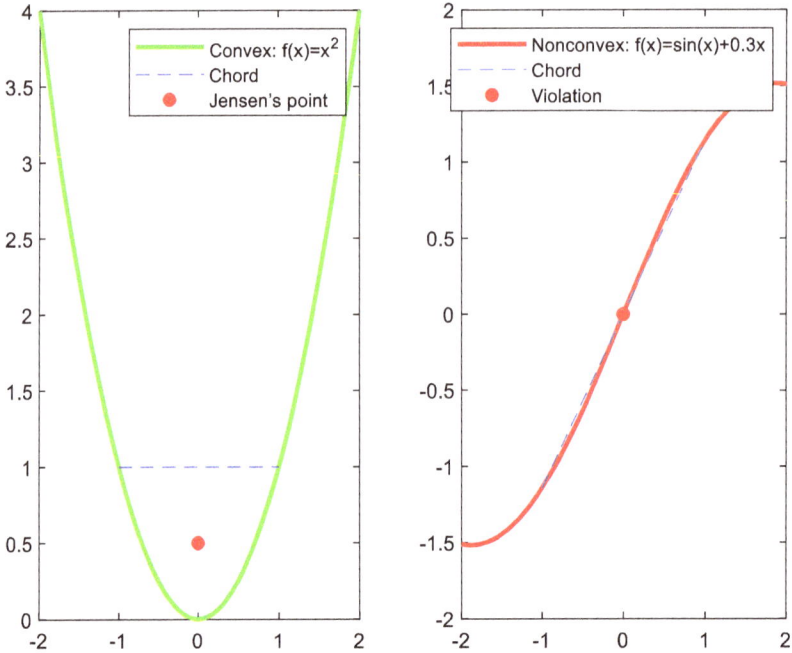

Figure 2.7: A convex function (left) and a non-convex function (right).

2.3.2 Tests for Convexity

For twice-continuously differentiable functions, convexity can be verified using the Hessian matrix. A function is convex if and only if $H(x)$ is *positive semidefinite* $(H(x) \succeq 0)$ for all x in its domain. Numerically, this requires that all eigenvalues of H are nonnegative. We can distinguish between strictly convex and convex, as shown in Table 2.2.

Table 2.2: Relationship between function and Hessian.

$f(x)$ is	$H(x)$ is	All eigenvalues of $H(x)$ are
Strictly convex	Positive definite	>0
Convex	Positive semidefinite	=>0
Strictly concave	Negative semidefinite	<=0
Strictly concave	Negative definite	<0

Here is an example. Consider a common substrate-inhibited reaction rate expression found in bioreactors and enzymatic systems:

$$r(C_A) = \frac{k_1 C_A}{1 + k_2 C_A^2}$$

where C_A is the substrate concentration (mol/L), k_1 the intrinsic rate constant (1/s), and k_2 the inhibition coefficient (L^2/mol^2). We want to determine whether this rate expression is convex to assess optimization challenges in reactor design.

We must compute the second derivative first. The second derivative has to be non-negative for all $C_A > 0$:

$$\frac{d^2 r}{dC_A^2} = \frac{2k_1 k_2 C_A \left(k_2 C_A^2 - 3\right)}{\left(1 + k_2 C_A^2\right)^3}$$

We can see that for all $C_A > 0$ the denominator is $\left(1 + k_2 C_A^2\right)^3 > 0$. In the numerator, the sign depends on $2k_1 k_2 C_A \left(k_2 C_A^2 - 3\right)$

The convex region (second derivative is positive) is: $k_2 C_A^2 - 3 \geq 0 \rightarrow C_A \geq \sqrt{3/k_2}$

The non-convex region (second derivative is negative): $C_A \leq \sqrt{3/k_2}$. The inflection point is at $C_A = \sqrt{3/k_2}$. We might derive from this that at low concentration, inhibition is negligible and that the rate increases nearly linearly. At high concentration, inhibition dominates, causing the rate to decrease.

For a numerical example, we could use: $k_1 = 1$ s^{-1} and $k_2 = 1$ L^2/mol^2. We will find that:

The inflection point: $C_A = \sqrt{3} \approx 1.732$ mol/L
The convex region is $0 < C_A < \frac{1}{732}$ mol/L
The non-convex region $C_A > 1.732$ mol/L
We can test with MATLAB:

```
syms CA k1 k2 positive;
r = k1 * CA / (1 + k2 * CA^2);
H = diff(r, CA, 2);
H_function = matlabFunction(H, 'Vars', {CA, k1, k2});

% Evaluate at CA = 1 (nonconvex) and CA = 2 (convex)
H_1 = H_function(1, 1, 1)  % Returns -0.5 (negative → nonconvex)
H_2 = H_function(2, 1, 1)  % Returns 0.056 (positive → convex)
```

2.3.3 Implications for Optimization

Convexity guarantees that any local minimum is a global minimum, fundamentally simplifying optimization. If we consider optimization problems with constraints, we distinguish between: 1) convex constraints + convex objective function (global optimum can be found via gradient-based methods, and 2) non-convexities, which might

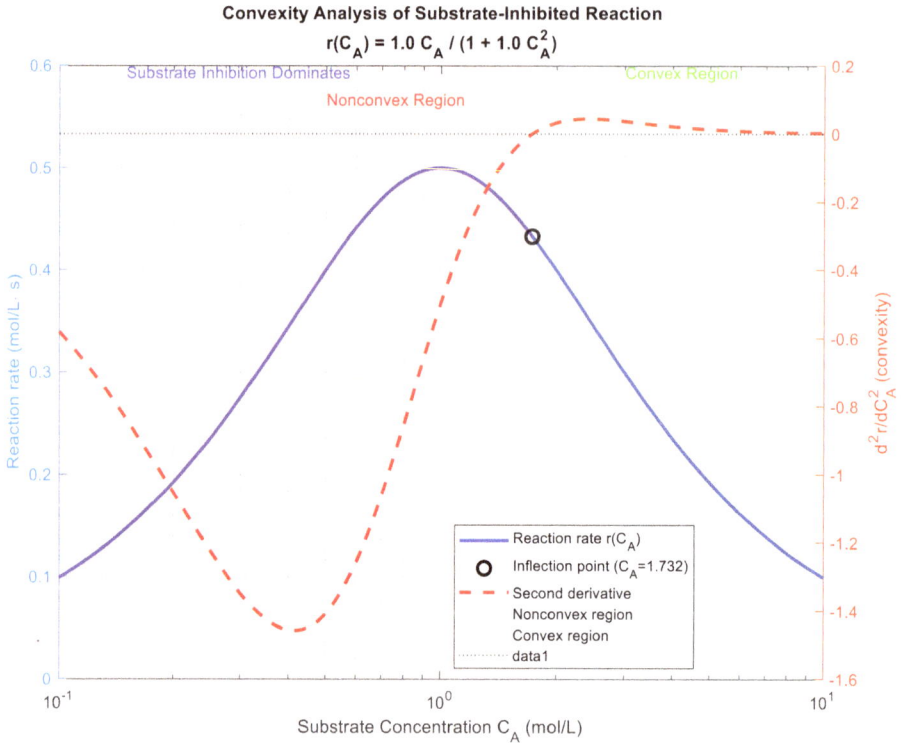

Figure 2.8: The reaction rate, second derivative, inflection point, and the convex and non-convex regions for different concentrations.

introduce multiple local optima, requiring global optimization techniques. We often encounter non-convexity in chemical processes, for example, from bilinear terms (e.g., $C_A F$ in reactor balances), discontinuous costs (fixed charges for equipment) or from phase transitions, where Gibbs energy landscapes are non-convex. Figure 2.8 shows the reaction rate, its second derivative, the inflection point and the convex and nonconvex regions for different concentrations.

2.4 Introduction to Numerical Methods

Numerical methods form the backbone of modern process optimization, enabling the solution of complex problems where analytical approaches are infeasible. This section introduces three fundamental numerical techniques: finite differences for gradient approximation, iterative methods for solving linear systems, and convergence analysis in optimization. Each topic is presented with theoretical foundations, practical considerations, and illustrative examples.

2.4.1 Finite Differences and Error Analysis

Derivatives play a central role in optimization, particularly in gradient-based methods. When analytical derivatives are unavailable – due to black-box simulations or computational complexity – finite differences provide a practical alternative. The simplest approximation is the *forward difference*:

$$f'(x) \approx \frac{f(x+h) - f(x)}{h}$$

where h is a small step size. While easy to implement, this method introduces a *truncation error* of order $O(h)$, meaning the error scales linearly with h. A more accurate alternative is the *central difference*:

$$f'(x) \approx \frac{f(x+h) - f(x-h)}{2h}$$

which reduces the truncation error to by symmetrically sampling points around x. However, this comes at the cost of an additional function evaluation.

The total error in finite differences arises from two sources 1) the truncation error, as introduced before – due to the approximation of a derivative via a Taylor series expansion, and 2) the *round-off error*, caused by finite-precision arithmetic, which becomes significant when is too small.

A practical rule for choosing h is $h \approx \sqrt{\epsilon}$, where is *machine epsilon* (approximately 10^{-8} for double-precision floating point arithmetic).

Below is a small example of a forward difference approximation in Python, to compute $f'(x)$ when $f(x) = x^2$:

```python
def forward_diff(f, x, h):
    """Forward difference approximation of f'(x)"""
    return (f(x + h) - f(x)) / h
# Example usage:
f = lambda x: x**2  # Test function
x0 = 1.0            # Point of evaluation
h = 1e-5            # Step size
derivative = forward_diff(f, x0, h)
print(f"Approximate derivative: {derivative}")
```

2.4.2 Iterative Methods for Linear Systems

Many optimization algorithms require solving linear systems of the form $Ax = b$. For large-scale problems (e.g., process flowsheets with thousands of variables), direct methods like Gaussian elimination become computationally expensive. Instead, iterative methods are preferred due to their memory efficiency and ability to exploit sparsity.

The *Jacobi method* decomposes the matrix A into its diagonal D and off-diagonal components $L + U$. The update rule is:

$$x^{k+1} = D^{-1}\left(b - (L + U)x^k\right)$$

Convergence is guaranteed if A is strictly diagonally dominant. An improvement over the Jacobi method is the *Gauss-Seidel method*, which uses the most recent updates with the same iteration:

$$x^{k+1} = (D + L)^{-1}\left(b - Ux^k\right)$$

This typically converges faster than Jacobi but requires careful handling for parallel implementations. Let us consider solving the steady-state 2D heat equation:

$$\nabla^2 T = 0$$

$$\nabla^2 T = \frac{\partial^2 T}{\partial x^2} + \frac{\partial^2 T}{\partial y^2}$$

The left wall ($x = 0$) has a fixed temperature $T = 100$ (*Dirichlet condition*). The other walls: $T = 0$. We can use central differences to approximate the derivatives:

$$\frac{T_{i+1,j} + T_{i-1,j} + T_{i,j+1} + T_{i,j-1} - 4T_{i,j}}{h^2} = 0$$

where we discretized the domain $[0,1] \times [0,1]$ into $N \times N$ grid points. The step size $h = 1/(N-1)$. From the finite differences we have obtained, effectively a linear system ($AT = b$). By simplifying, the Gauss–Seidel update rule becomes:

$$T_{i,j} = \frac{1}{4}\left(T_{i+1,j} + T_{i-1,j} + T_{i,j+1} + T_{i,j-1}\right)$$

This rule can be implemented in MATLAB:

```
% Grid and boundary conditions
N = 50; [X,Y] = meshgrid(linspace(0,1,N));
T = zeros(N); T(:,1) = 100; % Hot left wall

% Gauss-Seidel iteration
max_iter = 1e4; tol = 1e-6;
for k = 1:max_iter
    T_old = T;
    for i = 2:N-1
```

```
    for j = 2:N-1
        T(i,j) = 0.25*(T(i+1,j) + T(i-1,j) + T(i,j+1) + T(i,j-1));
    end
  end
  if norm(T - T_old, 'fro') < tol, break; end
end
surf(X,Y,T); title('Steady-State Temperature');
```

Figure 2.9 shows the calculated temperature profile.

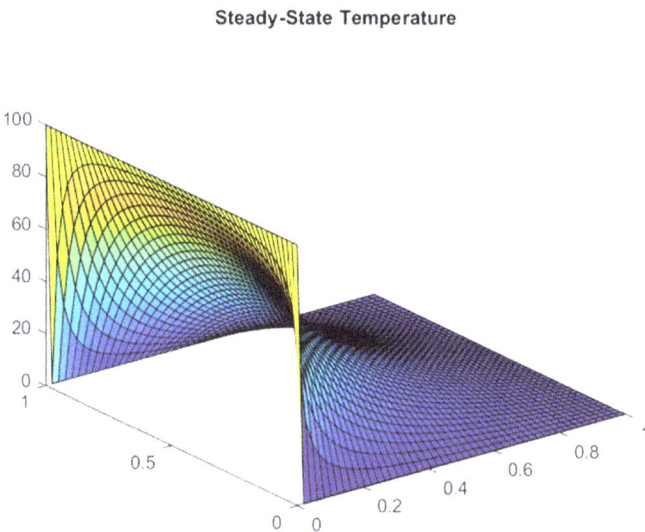

Figure 2.9: Computed temperature profile with Gauss–Seidel.

For a symmetric positive-definite (SPD) matrix A, the quadratic form is:

$$\phi(x) = \frac{1}{x}x^T A x - b^T x$$

It has a unique minimum, where its gradient is zero:

$$\nabla\phi(x) = Ax - b = 0$$

which gives the original system. In other words, minimizing $\phi(x)$ is the same as solving $Ax = b$. This concept is exploited in the *conjugate gradient* CG method. It is particularly efficient for large, sparse systems (common in optimization, CFD, and process modeling).

As an example:

Let $A = \begin{bmatrix} 2 & 1 \\ 1 & 2 \end{bmatrix}$ (SPD) and $b = \begin{bmatrix} 3 \\ 4 \end{bmatrix}$

The quadratic form is:

$$\phi(x) = \frac{1}{2}(2x_1^2 + 2x_1 x_2 + 2x_2^2 - 3x_1 - 4x_2)$$

Its minimum is at $x^* = A^{-1}b = \begin{bmatrix} \frac{2}{3}, \frac{5}{3} \end{bmatrix}^T$ which solves $Ax = b$

2.4.3 Newton's Method for Solving Nonlinear Equations

The core of process optimization involves solving nonlinear equations of the form $F(x) = 0$ where F and x are n-dimensional vectors. These arise from material and energy balances, equilibrium relationships (e.g., flash calculations and reaction kinetics), and the optimality conditions (Karush-Kuhn-Tucker conditions).

Newton's method, also known as the Newton-Raphson method, is the fundamental algorithm for this task. It is an iterative method that leverages local linearization to find successively better approximations to the root.

The core idea is to approximate the nonlinear function $F(x)$ by its first-order Taylor series expansion, around the current guess x_k:

$$F(x) \approx F(x_k) + J(x_k)(x - x_k)$$

where $J(x_k)$ is the Jacobian matrix of F evaluated at x_k. The Jacobian contains all the first-order partial derivatives:

$$J(x) = \begin{bmatrix} \frac{\partial F_1}{\partial x_1} & \cdots & \frac{\partial F_1}{\partial x_n} \\ \vdots & \ddots & \vdots \\ \frac{\partial F_n}{\partial x_1} & \cdots & \frac{\partial F_n}{\partial x_n} \end{bmatrix}$$

We want to find the next iterate such that this linear approximation is zero:

$$F(x_k) + J(x_k)(x_{k+1} - x_k) = 0$$

Solving for gives the Newton iteration formula:

$$x_{k+1} = x_k + [J(x_k)]^{-1}F(x_k)$$

In practice, we avoid calculating the inverse. Instead, we solve the equivalent linear system for the step δ_k at each iteration:

$$J(x_k)\delta_k = -F(x_k)$$

$$x_{k+1} = x_k + \delta_k$$

Algorithm steps:
1. Initialize: provide an initial guess x_0 and tolerance ε
2. Iteration:
   a. For $k = 1, 2, \ldots$
   b. Evaluate the function $F(x_k)$ and the Jacobian $J(x_k)$
   c. Solve the linear system $J(x_k)\delta_k = -F(x_k)$ for δ_k
   d. Update the solution: $x_{k+1} = x_k + \delta_k$
3. Termination: check for convergence; common criteria are:
   a. $\|\delta_k\| < \epsilon$
   b. $\|F(x_{k+1})\| < \epsilon$

Newton's method converges quadratically (the error squared at each step) near a root, making it exceptionally fast. However, this strong theoretical result comes with the caveats critical for practical implementation.
- Initial guess: convergence is only guaranteed if the initial guess is sufficiently close to the solution. A poor guess can lead to divergence;
- Jacobian evaluation: The need to compute the Jacobian at every step can be computationally expensive and analytically tedious;
- Singular Jacobian: If $J(x_k)$ becomes singular, the step δ_k cannot be computed.

These challenges can be addressed by:
- Providing good initial estimates (using physical insight or simpler models to start the algorithm);
- Using finite differences (if analytical derivatives are unavailable, the Jacobian can be approximated numerically);
- Employing Damped Newton's method. This is a modification where the step is scaled with a damping parameter $0 \le \lambda \le 1$, $x_{k+1} = x_k + \lambda\delta_k$ This helps stabilize convergence when the initial guess is poor;
- Using hybrid methods: algorithms like Powell's method or Broyden's method do not require the Jacobian.

Newton's method is directly extended to optimization (minimization) of a scalar function $f(x)$. Finding the optimum involves solving the nonlinear system $\nabla f(x) = 0$. Applying Newton's method to the gradient leads to the iteration:

$$x_{k+1} - [H(x_k)]^{-1}\nabla f(x_k)$$

where H is the Hessian with second derivatives. This demonstrates the foundational role of the method; it is the bridge between solving nonlinear equations and finding optima.

2.4.4 Convergence and Stability

The reliability of numerical methods hinges on their convergence and stability. Understanding these properties is crucial for selecting appropriate algorithms and diagnosing failures. Common criteria include:

1. Absolute error: $||x^{k+1} - x^k|| < \epsilon$

2. Relative error: $\frac{||x^{k+1} - x^k||}{||x^k||} < \epsilon$

3. Residual norm: $||Ax^k - b|| < \epsilon$

A solution procedure might also not lead to convergence or instability when the matrix A is ill-conditioned. A matrix A is ill-conditioned when its condition number $\kappa(A) = ||A||||A||^{-1}$ is large.

2.5 Takeaway

This chapter provides the essential analytical and numerical tools required to formulate, analyze, and solve optimization problems in chemical engineering. The chapter begins with a review of calculus concepts, emphasizing derivatives and gradients, which quantify how system outputs respond to input changes, as illustrated through the Arrhenius reaction rate model. It then extends these ideas to multivariable systems using partial derivatives and the Jacobian matrix, enabling sensitivity analysis for applications like reactor yield optimization. The discussion progresses to second-order derivatives via the Hessian matrix, which reveals curvature properties to classify critical points as minima, maxima, or saddle points – key for diagnosing optimization landscapes. Linear algebra fundamentals, including matrix operations, eigenvalues, and norms, are introduced to solve systems of equations arising in process flowsheets and stability analysis. A dedicated section on convexity explains how convex functions and sets simplify optimization by guaranteeing global optima, while non-convexities, such as those in substrate-inhibited reactions, pose challenges requiring advanced techniques. The chapter then transitions to numerical methods, covering finite differences for gradient approximation and iterative solvers like Jacobi, Gauss–Seidel, and conjugate gradient for large linear systems, demonstrated through a 2D heat equation example. Convergence criteria and ill-conditioning are analyzed to ensure solver reliability. By integrating theory with practical examples – from parameter estimation to computational fluid dynamics – the chapter equips readers to tackle optimization problems with rigor and efficiency.

2.6 Exercises

Exercise 1: Derivative Calculation ★
Compute the gradient ∇f for $f(x,y) = x^2 e^{-y} + y\ln(x)$ at $(x,y) = (1,0)$.
Hint: Use partial derivatives and verify units if applicable.

Exercise 2: Hessian and Critical Points ★★
Find the critical points and classify them (min/max/saddle) for $f(x,y) = x^3 + 3xy^2 - 3x$.
Hint: Compute the Hessian matrix and eigenvalues.

Exercise 3: Matrix Inversion ★
Solve $Ax = b$ for $A = [2112]$, $b = [34]$ using:
a) Direct inversion.
b) The conjugate gradient method (implement in Python/MATLAB).

Exercise 4: Eigenvalue Analysis ★★
Compute the eigenvalues of $A = [41;14]$ and discuss their implications for optimization stability.

Exercise 5: Convexity Test ★★
Show whether $f(x) = x_1^2 + x_2^2 - x_1 x_2$ is convex.
Hint: Check the Hessian's positive definiteness.

Exercise 6: Non-convex Case Study ★★★
Analyze the convexity of the reaction rate $r(C_A) = 2k_1 C_A/(1 + k_2 C_A)$ for $CA > 0$. Identify the inflection point.

Exercise 7: Finite Differences ★
Approximate $f'(x)$ for $f(x) = \sin(x)$ at $= \pi/4$ using forward, backward, and central differences (h = 0.1 $h = 0.1$). Compare errors.

Exercise 8: Gauss-Seidel Versus Jacobi ★★
Solve the 2D heat equation $\nabla^2 T = 0$ on a 5×5 grid with $T_{\text{left}} = 100, T_{\text{other}} = 0$. Implement both methods and compare convergence rates.

Further Reading

Boyd, Stephen, and Lieven Vandenberghe. 2004. *Convex Optimization*. Cambridge: Cambridge University Press.
Strang, Gilbert. 2016. *Linear Algebra and Its Applications*. 5th ed. Boston: Cengage.
Nocedal, Jorge, and Stephen J. Wright. 2006. *Numerical Optimization*. 2nd ed. New York: Springer.

LeVeque, Randall J. 2007. *Finite Difference Methods for Ordinary and Partial Differential Equations.* Philadelphia: SIAM.
Saad, Yousef. 2003. *Iterative Methods for Sparse Linear Systems*. 2nd ed. Philadelphia: SIAM.

Group Project: Optimization of a Chemical Reactor Design

Difficulty: ★★☆ (Intermediate)
Tools: Python (NumPy/SciPy), MATLAB, or Julia.

Project Overview

Design and optimize a continuous stirred-tank reactor (CSTR) for a hypothetical exothermic reaction, balancing reaction yield and cooling costs. Students will apply calculus, linear algebra, and numerical methods from Chapter 2 to solve the problem.

Project Tasks

1. **Problem Formulation**
   - **Objective**: Maximize profit = (Product Value × Yield) − (Cooling Cost).
   - **Constraints**:
     - Reaction kinetics: $r=k0e-EA/RTCA$ (Arrhenius equation).
     - Cooling cost: $Cost=a(T-Tcoolant)$.
     - Safety: $T\leq400K$.
2. **Mathematical Modeling**
   - Derive the **gradient** of profit w.r.t. temperature T and concentration CA.
   - Compute the **Hessian** to check convexity of the profit function.
   - Discretize the system into a grid for numerical analysis.
3. **Numerical Optimization**
   - Implement **finite differences** to approximate gradients (forward/central differences).
   - Use **Gauss-Seidel** to solve the temperature distribution in the reactor.
   - Apply **conjugate gradient** to optimize T and CA iteratively.
4. **Analysis and Report**
   - Compare analytical versus numerical results.
   - Visualize trade-offs (yield vs. cost) and identify the optimal (T,CA).
   - Present findings in a 10-slide PowerPoint or 5-page report.

Deliverables

1. **Code**: Python/MATLAB scripts for gradient/Hessian calculations and solvers.
2. **Report**: Summary of methods, results, and sensitivity analysis (e.g., how α affects optimum).
3. **Presentation**: 10-min pitch on key insights.

CSTR Optimization Project Data Table

Parameter	Symbol	Value	Units	Notes
Reaction kinetics				
Pre-exponential factor	k_0	5.0×1,075.0×107	s^{-1}	From Arrhenius equation
Activation energy	E_A	65,000	J/mol	Typical for exothermic reactions
Gas constant	R	8.314		Universal constant
Feed concentration	C_{A0}	2.0	Lmol/L	Initial concentration of reactant A
Operating conditions				
Temperature range	T	300–400	K	Safety constraint
Coolant temperature	$T_{coolant}$	290	K	Fixed cooling utility temperature
Cooling cost coefficient	α	0.02		Cost per unit temperature difference
Economic Parameters				
Product value	–	10	$/mol	Revenue per mole of product
Operating time	–	105,105	s	~1 day of continuous operation
Reactor geometry				
Volume	V	1.0	m^3	Standard lab-scale CSTR
Flow rate	Q	0.1	m^3/s	Residence time $\tau=V/Q$

3 Formulating Optimization Problems

The formulation of a problem is more essential than its solution. – Albert Einstein

3.1 Introduction

Einstein's quote underscores the foundational importance of clearly defining optimization challenges before attempting to solve them.

The BP Texas City Refinery Disaster (2005)

What Failed?

A distillation column in the isomerization (ISOM) unit overflowed during startup due to:

- **Missing level/pressure constraints** in the operating procedure.
- **Disabled alarms** that ignored high-level warnings.
- **No automatic shutdown** for overfill scenarios.

Optimization Link

1. **Flawed Formulation**:
   - Objectives prioritized throughput over safety.
   - No inequality constraints for max liquid level (e.g., *Level ≤ 100%*).
2. **Catastrophic Result**:
   - Overfill → Pressure surge → Explosion → **15 fatalities**.

Lessons for Engineers

✓ **Hard Constraints Are Non-Negotiable**: Always include safety limits (pressure, temperature, level).
✓ **Test Edge Cases**: Simulate startups/shutdowns in formulations.
✓ **Balance Objectives**: Efficiency ≠ Safety; use multi-objective optimization (Ch. 7).
Source: U.S. Chemical Safety Board (CSB) Report, 2007.

To illustrate the consequences of poor formulation, consider a real-world example where a chemical plant experienced a catastrophic failure due to an overlooked pressure constraint in its reactor design. Such incidents highlight how even advanced solution methods can fail if the problem is improperly structured. In 2005, the BP Texas City disaster – caused by an overfilled distillation column during startup – highlighted the lethal consequences of overlooking constraints. While not a reactor, this case illustrates how poor optimization formulations (e.g., missing pressure or level limits) can lead to catastrophic outcomes. Figure 3.1. shows the catastrophic outcomes of the BP dissaster. This chapter teaches how to avoid such errors by rigorously defining constraints and variables.

The primary goal of this chapter is to equip readers with a systematic approach to formulating optimization problems. By mastering this skill, common pitfalls, such as infeasible constraints or mismatched objectives, can be avoided, and lay the groundwork for applying solution techniques covered in later chapters.

https://doi.org/10.1515/9783111342283-003

Figure 3.1: The BP Texas City Refinery disaster.

3.2 Components of an Optimization Problem

Every optimization problem comprises three fundamental elements: the *objective function*, *decision variables*, and *constraints*. Understanding these components is crucial for formulating effective problems.

3.2.1 Objective Function

The objective function quantifies the goal of the optimization, whether it involves minimizing costs, maximizing efficiency, or balancing competing priorities. For instance, in a chemical process, objectives might include minimizing energy consumption or maximizing product yield. Economic objectives, such as net present value (NPV), often compete with technical goals like purity or conversion rates.

3.2.2 Decision Variables

Decision variables are the adjustable parameters that influence the objective function. These can be continuous (e.g., flow rates, temperatures), discrete (e.g., the number of reactors), or binary (e.g., yes/no decisions for equipment installation). Selecting the right variables requires ensuring they are independent and actionable. For example, in a distillation column optimization, the reflux ratio and feed temperature are typical decision variables.

3.2.3 Constraints

Constraints define the boundaries within which the solution must reside. Equality constraints, such as mass or energy balances, enforce strict relationships between variables. Inequality constraints, like safety limits (e.g., temperature ≤ 500 K) or capacity restrictions, bound the feasible region. Constraints can be "hard" (non-negotiable, such as regulatory requirements) or "soft" (preferential, such as target purity levels).

To solidify these concepts, consider the following comparative example: *In a reactor optimization problem, the objective might be to maximize yield, with decision variables including inlet concentration and temperature. Constraints could involve thermal stability (e.g., temperature ≤ 450 K) and budget limits (e.g., catalyst cost ≤ $10,000). In contrast, a distillation column optimization might prioritize minimizing energy use, with variables like reflux ratio and constraints such as product purity ≥ 98%.*

The problem can be mathematically written as follows:

$\min f(x, d)$	–	The objective function
Subject to:		
$g(x, d) = 0$	–	The equality constraints
$h(x, d) > 0$	–	The inequality constraints
$x^L \le x \le x^U$	–	The bound constraints
$x \in X, \ d \in D$	–	The domain over which the decisions are defined

Figure 3.2 shows how PhD student Henrique is combining the basic ingredients of an optimization problem: objective, decisions and constraints. Depending on the nature of the objective function and constraints, we speak of a linear program (LP, objective and constraints are linear, all variables are continuous), a *nonlinear program* (NLP, objective and/or constraints are nonlinear, all variables are continuous), a mixed-integer linear program (MILP, objective and constraints are linear, but continuous and discrete variables occur), and a *mixed-integer nonlinear program* (MINLP, objective and/or constraints are nonlinear, but continuous and discrete variables occur). A special MINLP is the *mixed-integer quadratic program* (MIQP, where the nonlinear objective/constraints are quadratic). There are also problems with only discrete variables, which are often denoted as integer programs (IPs).

You notice that we speak of programs and programming. Actually, within the field of optimization, the art of setting up and solving optimization models is called *mathematical programming.* Mathematical programming originated in the 1940s–1950s alongside the development of optimization theory. Particularly in the context of linear programming, the word programming derives from military logistics and economic planning, where it referred to systematic decision-making. George Dantzig, the father of linear programming, used the term in 1947 in his work on the Simplex method for the US Air Force's planning problems.

"The term reflects its roots in WWII-era planning – long before 'programming' meant writing code." – **George Dantzig, *Linear Programming and Extensions* (1963)**

After setting the goals for a project and collecting data, the process of formulating an optimization problem can start. Let us look here at an example:

3.2.4 The Refinery Blending Problem

Alkylate, cat-cracked gasoline, and straight-run gasoline are blended to make aviation gasolines A and B and two grades of motor gasoline. The specifications for motor gasoline are not as rigid as for aviation gas. Physical property and production data for the inlet streams are as follows:

RVP, Reid vapor pressure (measure of volatility); ON, octane number and the number in parenthesis shows the number of mL/gal of tetraethyl lead (TEL).

Table 3.1: Crude data.

Stream	RVP	ON(0)	ON(4)	Available (bbl/day)
Alkylate	5	94	108	4,000
Cat-cracked gasoline	8	84	94	2,500
Straight-run gasoline	4	74	86	4,000

Table 3.2: Data for blended products.

Product	RVP	TEL level	ON	Profit ($/bbl)
Aviation gasoline A	≤7	0	= > 80	5.00
Aviation gasoline B	≤7	4	= > 91	5.50
Leaded motor gasoline	–	4	= > 87	4.50
Unleaded motor gasoline	–	0	= > 91	4.50

The goal here is to translate this information into a linear program. We need to think of objective, decisions, and constraints.

Objective: from the table, we find data on how much profit a barrel of product can yield in dollars per barrel. This implies that we are dealing with *a profit maximization problem.*

The objective function could look like:

$$\max P = 5.00Y_1 + 5.50Y_2 + 4.50Y_3 + 4.50Y_4$$

where Y_1 is the amount of produced aviation fuel A (bbl), Y_2 is the amount of aviation fuel B (bbl), Y_3 is the amount of leaded motor gasoline (bbl), and Y_4 is the amount of unleaded motor gasoline (bbl). We could use Y_j instead, where j is a set containing aviation fuel A (1), aviation fuel B (2), leaded gasoline (3), and unleaded gasoline (4).

The Y's are decision variables, but they are not free; they depend on the amounts of crude that are used to blend them.

Decisions: We can introduce a decision variable $X_{i,j}$ which denotes the amount of crude i that is used to make a product j. i is a set containing alkylate (1), cat-cracked gasoline (2), and straight-run gasoline (3).

Lastly, we have to think of the *constraints*. Provided the data above, we will have the following constraints:
1. Mass balance
2. Capacity limits of the crudes
3. Quality criteria (in terms of octane number and vapor pressure).

The general *mass balance*:

$$Y_j = \sum_i X_{i,j} \ \forall j$$

So for the aviation fuel A:

$$Y_1 = X_{11} + X_{21} + X_{31}$$

For aviation fuel B:

$$Y_2 = X_{12} + X_{22} + X_{32}$$

For leaded gasoline

$$Y_3 = X_{13} + X_{23} + X_{33}$$

For unleaded gasoline:

$$Y_4 = X_{14} + X_{24} + X_{34}$$

We also have to make sure that we do not exceed the inventory *capacity* of the crudes:

$$\sum_j X_{ij} \leq b_i \forall i$$

where b_i is the available amount of crude i. We could write the capacity constraint per crude:

For alkylate:

$$X_{11} + X_{12} + X_{13} + X_{14} \leq 4,000$$

For cat-cracked crude:

$$X_{21} + X_{22} + X_{23} + X_{24} \leq 2,500$$

For straight crude:

$$X_{31} + X_{32} + X_{33} + X_{34} \leq 4,000$$

We also have to ensure that the products have the right *quality*, in terms of octane number and vapor pressure. No detailed information is provided regarding blending behavior, so we assume linear blending behavior. For the octane number:

$$ON_j\, Y_j \leq \sum_i ON_i\, X_{i,j} \;\forall j$$

where ON_j is the required octane number of product j and where ON_i is the octane number of crude i. Make sure to pick the octane numbers with the right TEL level. We can write the constraint for each product.

So for the aviation fuel A:

$$80Y_1 \leq 94X_{11} + 84X_{21} + 74X_{31}$$

For aviation fuel B:

$$91Y_2 \leq 108X_{12} + 94X_{22} + 86X_{32}$$

For leaded gasoline

$$87Y_3 \leq 108X_{13} + 94X_{23} + 86X_{33}$$

For unleaded gasoline:

$$91Y_4 \leq 94X_{14} + 84X_{24} + 74X_{34}$$

We can write two additional quality constraints for the RVP. For aviation fuel A:

$$7Y_1 \geq 5X_{11} + 8X_{21} + 4$$

and for aviation fuel B:

$$7Y_2 \leq 5X_{12} + 8X_{22} + 4X_{32}$$

The model is now complete. We have developed a linear program, with 13 constraints and 16 variables (of which 12 are free variables). We can now use an optimization software package to code and solve this model. Below you can find a GAMS implementation:

```
$title REFINERY
* This is a model to optimize keresone upgrading

variables
Z overal profit

positive variables
X11 alkylate for gasoline A
X21 cat cracked for gasoline A
X31 straight for gasoline A
X12 alkylate for gasoline B
X22 cat cracked for gasoline B
X32 straight for gasoline B
X13 alkylate for leaded
X23 cat cracked for leaded
X33 straight for leaded
X14 alyklate for unleaded
X24 cat cracked for unleaded
X34 straight for unleaded
Y1 amount of gasoline A
Y2 amount of gasoline B
Y3 amount of leaded motor oil
Y4 amount of unleaded motor oil

equations
profit overall profit
capX1 capacity of X1
capX2 capacity of X2
capX3 capacity of X3
massP1 mass balance P1
massP2 mass balance P2
massP3 mass balance P3
massP4 mass balance P4
ONY1 octane for Y1
ONY2 octane for Y2
ONY3 octane for Y3
ONY4 octaine for Y4
RVPY1 rvp for Y1
RVPY2 rvp for Y2;
profit .. Z =e= 5*Y1 + 5.5*Y2 +4.5*Y3 + 4.5*Y4;
```

```
capX1 .. X11 + X12 + X13 + X14 =l= 4,000;
capX2 .. X21 + X22 + X23 + X24 =l= 2,500;
capX3 .. X31 + X32 + X33 + X34 =l= 4,000;
massP1 .. X11+X21+X31 =e=Y1;
massP2 .. X12+X22+X32 =e=Y2;
massP3 .. X13+X23+X33 =e=Y3;
massP4 .. X14+X24+X34 =e=Y4;
ONY1 .. 94*X11 + 84*X21 + 74*X31 =g= 80*Y1;
ONY2 .. 108*X12 + 94*X22 + 86*X32 =g= 91*Y2;
ONY3 .. 108*X13 + 94*X23 + 86*X33 =g= 87*Y3;
ONY4 .. 94*X14 + 84*X24 + 74*X34 =g= 91*Y4;
RVPY1 ..5*X11 + 8*X21 + 4*X31 =l= 7*Y1;
RVPY2 ..5*X12 + 8*X22 + 4*X32 =l= 7*Y2;
model REFINERY /all/;
solve REFINERY using LP maximizing Z;
```

Figure 3.2: PhD candidate Henrique has a mission to optimize.

3.3 Step-by-Step Problem Formulation

Formulating an optimization problem requires a structured approach to ensure completeness and accuracy. The following steps provide a roadmap for translating real-world challenges into mathematical models.

3.3.1 Step 1: Define the System

Begin by delineating the system boundaries. For example, in a heat exchanger network, decide whether to include pumps or auxiliary cooling units. A block flow diagram can help visualize the system and identify key components.

3.3.2 Step 2: Identify Goals

Clearly articulate the primary and secondary objectives. For instance, a biodiesel production process might prioritize minimizing production cost, while secondary goals could include reducing greenhouse gas emissions.

3.3.3 Step 3: Select Decision Variables

Choose variables that directly impact the objectives and are feasible to adjust. A degrees-of-freedom analysis ensures the problem is neither under- nor overspecified. For a continuous stirred-tank reactor (CSTR), variables might include volume, flow rate, and temperature.

3.3.4 Step 4: Model Relationships

Establish mathematical relationships between variables. Linear relationships, such as pump power being proportional to flow rate, simplify the problem. Nonlinearities, such as reaction kinetics described by the Arrhenius equation, require more sophisticated treatment.

3.3.5 Step 5: Specify Constraints

List all constraints, ensuring they are necessary and non-redundant. For example, in a pipeline design, constraints might include maximum pressure ratings and minimum flow velocities to prevent erosion.

3.3.6 Step 6: Validate the Formulation

Test the formulation for feasibility and consistency. A quick check with extreme values (e.g., zero flow rate) can reveal hidden flaws.

FROM THE EXPERT: **Prof. Daniel Lewin**
Department of Chemical Engineering, Technion IIT, Haifa, Israel

Some Thoughts on the Role of Optimization in Process Systems Engineering (PSE)
"If the ideal process being designed was like a map of Scotland, the economic performance of well-designed process, when plotted against the key design parameters, should look like the lowlands of Scotland, that is rather insensitive to parameter values, rather than the highlands, with sharp peaks and deep valleys."

I have had a long academic career in which my educational activities span virtually all areas of process systems engineering (PSE). PSE is the branch of the chemical engineering discipline that exploits computational methods and tools for the analysis, design, control, optimization and effective operation of processing systems, and the design of products, across different scales and dimensions. I have taught courses in process and plant design, process control, process simulation, and numerical methods, among others. Here is a summary of the lessons learned regarding modeling and optimization activities in those areas:

Process Design. This is generally taught in a systematic linear approach for a reason. One first teaches students to close material and energy balances for a base case design's flowsheet. Then, by first employing heuristics and later systematic sequencing methods, the economic performance of the base case is improved. Next, by implementing heat and mass integration, the economic performance of the design is improved still further. Economic considerations should be addressed and used to drive engineering decisions at each step along the way. The case study that appears in Chapter 27 of Seider et al. [1] is particularly instructive. This describes the steps involved in completing a design project for the production of ammonia intended to be addressed by undergraduates. More specifically, the project development steps involve a continuous reassessment of the economic viability of the developing process as each new modification of the flowsheet is considered. Thus, economic considerations indeed drive the design decisions made along the way.

Process Control. In what concerns optimization, the principal role of process control is to ensure the desired optimal operating state of a production system be held as desired irrespective of possible disturbances. This is true even when the desired operating state may change periodically according to circumstances. Furthermore, an important purpose of feedback control is to compensate for uncertainties in the process models used to optimize the process. Indeed, depending on the desired performance, a successful controller could be designed on the basis of a crude process model. For example, to design the ubiquitous proportional-integral (PI) controller that can guarantee that most processes can be stably maintained at their setpoints, it is enough to know the sign of the static gain of the effect of the manipulated variable on the output variable. The seminal work of Manfred Morari and his students [2] formulated the relationship between potential closed-loop performance and the quality of the inherent process model.

Numerical Methods. A comprehensive course in numerical methods for chemical engineers needs to provide the basic building blocks needed to provide numerical approximations to the solutions of mathematical formulations that result from PSE activities [3, 4]. Such a course begins by instruction on the basic building blocks such as discrete formulation of differential operators and the numerical solution of linear and nonlinear equations. One then proceeds to applications such as the minimization of functions. Optimization inherently requires the minimization or maximization of an objective function formulated to achieve a given engineering objective, for example, the minimization of annual operating costs of a process or the maximization of its annual profits. Usually, the desired objective is constrained by physical and economic constraints, requiring the usage of constrained optimization methods such as linear programming (LP) and nonlinear programming (NLP). With what concerns process design, these methods can be helpful in improving the economic performance but are best used after sensitivity analysis has identified the design's parameters that most effect the result, and optimization is used to determine values for those key parameters.

References

[1]   Seider WD, Lewin DR, Seader JD, Widagdo S, Gani R, and Ng KM. "Product and Process Design Principles," 4th Ed., John Wiley and Sons, Hoboken NJ (2017)
[2]   Rivera DE, Morari M, Skogestad S, "Internal Model Control. 4. Controller Design," *Ind. Eng. Chem. Process Des. Dev.*, 25, 252–265 (1986)
[3]   Lewin DR and Barzilai A, "A Hybrid-Flipped Course in Numerical Methods for Chemical Engineers," *Comput. Chem. Eng.*, 172, 108167 (2023).
[4]   Zondervan E, "A Numerical Primer for the Chemical Engineer," 2nd Ed., CRC Press (2019).

3.4 Practical Techniques for Effective Formulation

Beyond the basics, several techniques enhance the robustness and solvability of optimization problems.

Dimensional Analysis: Ensuring all variables and constants share consistent units prevents errors. For example, converting all energy terms to joules avoids mismatches in calculations.

Variable Scaling: Normalizing variables (e.g., expressing temperature as $T' = T/100$ K) improves numerical stability in solvers. Poorly scaled problems, where variables range from 10^{-3} to 10^6, often fail to converge.

Simplifying Complex Problems: Breaking large problems into smaller subproblems (decomposition) or using surrogate models for computationally intensive functions can make optimization tractable.

Handling Uncertainty: Real-world problems often involve uncertain parameters (e.g., fluctuating feedstock prices). Robust optimization techniques, which account for variability, will be explored in Chapter 10.

"Always test your formulation with a trivial case – such as setting all inputs to zero – to catch errors early."

3.5 Case Study: Optimizing a Reactor System

To demonstrate the formulation process, this section walks through a detailed case study.

3.5.1 Problem Statement

A chemical plant aims to **maximize the production rate** of a desired product BB in a CSTR where the following reaction occurs:

$$A \rightarrow B$$

The reaction is exothermic and temperature-sensitive. The system must balance **yield**, **safety**, and **cost** under the following conditions:

- **Feed**: Pure A at concentration $C_{A0} = 2.5$ mol/L and flow rate $F = 100$ L/min.
- **Reactor Volume**: Fixed at $V = 500$ L.

Reaction Kinetics: First-order, with rate constant $k = k_0 \exp\left(-\frac{E_A}{RT}\right)$ ($k_0 = 5 \times 10^7$ min^{-1}), $E_A = 50$ kJ/mol.

Objectives and Challenges

The objective is to maximize the outlet concentration of B (C_B), to avoid thermal run-away ($T \leq 400$ K) and to minimize cooling costs (energy use $\propto T^2$).

3.5.2 Formulating the Optimization Problem

Step 1: Define the Objective Function

Maximize the production rate of B:

$$\max C_B = C_{A0} - C_A$$

where C_A is the outlet concentration of A.

Step 2: Identify Decision Variables

The primary variable is the reactor temperature T (directly impacts k). The secondary variable is the cooling rate Q (linked to the energy cost).

Step 3: Model Relationships

We set up the mass balance:

$$C_A = \frac{C_{A0}}{1 + k\tau}$$

Where $\tau = V/F$ is the residence time.
We set up the reaction rate:

$$k = 5 \times 10^7 \exp\left(-\frac{6,000}{T}\right)$$

Step 4: Specify Constraints

We have a safety constraint (thermal stability limit):

$$T \leq 400 \ K$$

We must ensure physical feasibility (ambient cooling limit)

$$T \geq 300 \ K$$

And we know that there are cooling costs, so we want to prevent excessive cooling:

$$\text{Cooling costs} \leq 500 \frac{\$}{\text{day}} (Q \leq 200 \ \text{kW})$$

3.5.3 Solution Methods

Option 1: Graphical Solution (Two Variables)

We plot C_B versus T for $Q = 0$ to 200 kW. The feasible region is bounded by the temperature constraints: $300 \leq T \leq 400$ K. We can now read the optimum at: $T = 319.9$ K, $Q = 198.4$ kW, yielding $C_B = 1.58$ mol/L (see Figure 3.3).

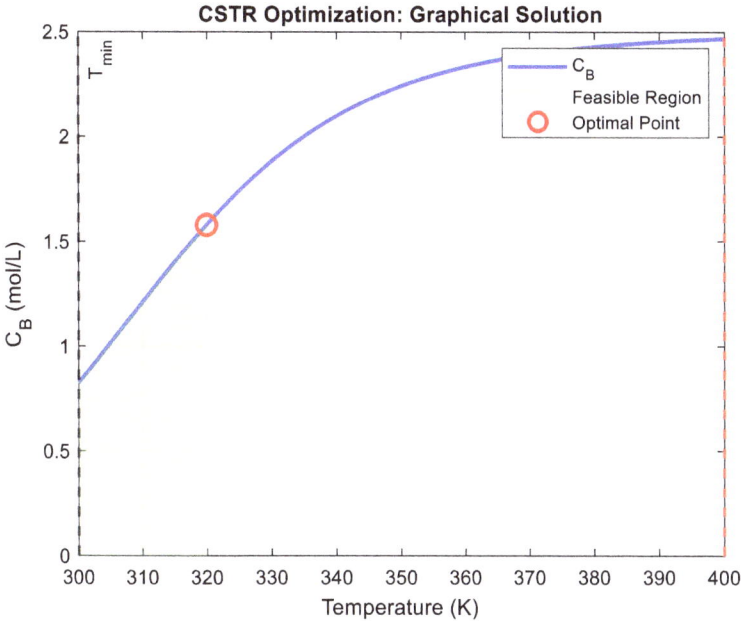

Figure 3.3: Reactor optimization example, graphical solution method.

Option 2: Numerical Optimization (MATLAB)

We can also use an optimization solver in MATLAB to find the optimum. Below is an implementation with the `fmincon` solver:

```
function cstr_optimization()
    % Parameters
    C_A0 = 2.5;      % Feed concentration (mol/L)
    F = 100;         % Flow rate (L/min)
    V = 500;         % Reactor volume (L)
    tau = V / F;     % Residence time (min)
    k0 = 5e4;        % Pre-exponential factor (min^-1)
    Ea = 50e3;       % Activation energy (J/mol)
    R = 8.314;       % Gas constant (J/mol·K)
    Q_max = 200;     % Maximum cooling (kW)

    % Graphical Analysis (C_B vs. T)
    T = linspace(300, 400, 1,000);   % Temperature range (K)
    k = k0 * exp(-Ea./(R * T));      % Reaction rate
    C_A = C_A0 ./ (1 + k * tau);     % Outlet A concentration
    C_B = C_A0 - C_A;                % Outlet B concentration
```

```
Q = 0.5 * (T - 300).^2;          % Cooling cost model

% Find feasible region
feasible = (T <= 400) & (Q <= Q_max);
[C_B_max, idx] = max(C_B(feasible));
T_opt_graphical = T(feasible);
T_opt_graphical = T_opt_graphical(idx);

% Plot results
figure;
plot(T, C_B, 'b-', 'LineWidth', 2); hold on;
fill([T(feasible),    fliplr(T(feasible))],    [C_B(feasible),
zeros(size(C_B(feasible)))], ...
    'g', 'FaceAlpha', 0.2, 'EdgeColor', 'none');
plot(T_opt_graphical, C_B_max, 'ro', 'MarkerSize', 10,
'LineWidth', 2);
xline(400, 'r--', 'T_{max}', 'LineWidth', 1.5);
xline(300, 'k--', 'T_{min}', 'LineWidth', 1.5);
xlabel('Temperature (K)'); ylabel('C_B (mol/L)');
title('CSTR Optimization: Graphical Solution');
legend('C_B', 'Feasible Region', 'Optimal Point', 'Location',
'northeast');
grid on;

% Numerical Optimization (fmincon)
fun = @(T) -objective(T, C_A0, k0, Ea, R, tau); % Negative for
maximization
T0 = 350; % Initial guess (K)A = []; b = []; Aeq = []; beq = []; % No
linear constraints
lb = 300; ub = 400; % Bounds
nonlcon = @(T) cooling_constraint(T, Q_max); % Nonlinear constraint
options = optimoptions('fmincon', 'Display', 'iter');
[T_opt, fval] = fmincon(fun, T0, A, b, Aeq, beq, lb, ub, nonlcon,
options);
C_B_opt = -fval; % Revert negative sign

% Display results
fprintf('Graphical Solution:\n');
fprintf(' Optimal T = %.1f K, C_B = %.2f mol/L\n', T_opt_graphical,
C_B_max);
fprintf('Numerical Solution (fmincon):\n');
```

```
    fprintf(' Optimal T = %.1f K, C_B = %.2f mol/L\n', T_opt, C_B_opt);
end


% Objective function (C_B to maximize)
function C_B = objective(T, C_A0, k0, Ea, R, tau)
    k = k0 * exp(-Ea/(R * T));
    C_A = C_A0 / (1 + k * tau);
    C_B = C_A0 - C_A;
end


% Nonlinear constraint (Q ≤ Q_max)
function [c, ceq] = cooling_constraint(T, Q_max)
    c = 0.5 * (T - 300)^2 - Q_max; % c ≤ 0
    ceq = []; % No equality constraints
end
```

Also `fmincon` con finds the same optimum as with the graphical method.

3.6 Common Mistakes and Remedies

Even experienced engineers encounter pitfalls during problem formulation. Recognizing these early can save time and resources.

Overlooking Constraints: Example: Ignoring pressure drops in a pipeline design can lead to undersized pumps and operational failures. Always cross-check with physical laws and operational limits.

Incorrect Linearity Assumptions: Assuming linear relationships for inherently nonlinear phenomena (e.g., reaction rates) leads to inaccurate models. Validate assumptions with experimental data or literature.

Poor Scaling: Solvers may struggle with variables differing by orders of magnitude. Rescale variables to similar ranges (e.g., 0.1 to 10) for reliable performance.

Ignoring Degrees of Freedom: Ensure the number of independent variables matches the problem's complexity. Over-constrained problems (e.g., more equations than variables) are often infeasible.

3.7 Tools and Software for Formulation

Practical implementation requires leveraging computational tools.

Equation-Based Tools: Software like GAMS and AMPL allows users to define problems using mathematical notation, ideal for complex systems.

Spreadsheet Solvers: Excel's Solver add-in provides an accessible platform for small-scale problems. For example, it can optimize a reactor's operating conditions using the formulated constraints.

3.8 Takeaway

This chapter emphasized the art and science of formulating optimization problems. Key takeaways include: 1) A well-defined objective function guides the entire optimization process, 2) Careful selection of decision variables and constraints ensures feasible and meaningful solutions, and 3) practical Practical techniques – scaling, decomposition, and validation – enhance problem solvability. The next chapter, *Linear Programming Methods*, will build on this foundation by introducing solution techniques for formulated problems.

Further Reading

Biegler, L. T. (2010). *Nonlinear programming: Concepts, algorithms, and applications to chemical processes.* Society for Industrial and Applied Mathematics.
Crowl, D. A., & Louvar, J. F. (2019). *Chemical process safety: Fundamentals with applications* (4th ed.). Pearson.
Rangaiah, G. P., & Bonilla-Petriciolet, A. (Eds.). (2013). *Multi-objective optimization in chemical engineering: Developments and applications.* Wiley.
Edgar, T. F., Himmelblau, D. M., & Lasdon, L. S. (2001). *Optimization of chemical processes* (2nd ed.). McGraw-Hill.
Biegler, L. T., Grossmann, I. E., & Westerberg, A. W. (1997). *Systematic methods of chemical process design.* Prentice Hall.

3.9 Exercises

Exercise 1: Basic LP Formulation ★
Problem: A chemical plant produces two products, X and Y, with profits of \$50/ton and \$80/ton, respectively. The production is constrained by:
- Raw material: 5 tons/day (each ton of X uses 2 tons; each ton of Y uses 1 ton).
- Labor: 8 h/day (each ton of X requires 1 h; each ton of Y requires 2 h).

Tasks:
1. Formulate the objective function and constraints.
2. Identify the decision variables.
3. Solve graphically (sketch the feasible region and optimal point).

Exercise 2: NLP with Safety Constraints ★★★

Problem: A reactor's yield (Y) depends on temperature (T) as:

$Y = 0.1T2 - 5T + 100$ ($300 \le T \le 400K$)

Safety requires:
- $T \le 380$ K (to prevent runaway reactions).
- Cooling cost $= \$0.2^*(T - 300)_2 \le \$500/day$.

Tasks:
1. Formulate the NLP to maximize yield.
2. Use calculus to find the unconstrained optimum.
3. Check if the unconstrained optimum violates safety/cost constraints.

Exercise 3: Integer Programming for Equipment Selection ★★★★

Problem: A plant must choose between three reactors ($R1, R2, R3$) with:
- Costs: $100k, $150k, $200k.
- Capacities: 5 tons/h, 8 tons/h, 10 tons/h.
- The plant needs ≥15 tons/h total capacity and can buy at most 2 reactors.

Tasks:
1. Define binary decision variables.
2. Formulate the MILP to minimize cost.
3. Propose a feasible solution (no solver required).

Exercise 4: Multi-objective Trade-Offs ★★★★★

Problem: A distillation column optimization has two objectives:
1. Maximize purity ($P = 90 + 0.5x$, where $^*x^*$ is reflux ratio).
2. Minimize energy cost ($C = 100 + 10x$).

Constraints:
- $1 \le {}^*x^* \le 10$.
- Purity must be ≥95%.

Tasks:
1. Plot the Pareto frontier (purity vs. cost).
2. Identify two feasible solutions that prioritize purity vs. cost.
3. Discuss trade-offs for a plant manager.

Group Project: Optimizing a Biofuel Production Process
Duration: 1 week (4–5 h of teamwork)
Group Size: 4 students
Difficulty: ★★★★ (Intermediate)

Learning Objectives:
–   Formulate a multi-objective optimization problem with economic and environmental constraints.
–   Apply LP/NLP techniques to a real-world scenario.
–   Collaborate on data analysis, modeling, and decision-making.

Project Scenario:
A startup aims to produce biofuel from algae in a photobioreactor. The team must optimize production while balancing cost, yield, and sustainability.

Key Parameters:
–   **Inputs:**
    –   **Nutrient A**: $20/kg (max 50 kg/day).
    –   **Nutrient B**: $30/kg (max 30 kg/day).
    –   **CO_2 Supply**: $10/kg (unlimited).
–   **Reaction:**
    –   Yield (grams of biofuel per day):

$$Y = 50x_1 + 70x_2 - 0.1x_1^2 - 0.2x_2^2$$

where x_1 = kg of nutrient A, x_2 = kg of nutrient B.
–   **Constraints:**
    –   **Budget**: ≤$1,500/day.
    –   **Carbon Footprint**: ≤100 kg CO_2/day (each kg of nutrient A emits 2 kg CO_2; each kg of Nutrient B emits 3 kg CO_2).

Tasks for the Team

1. **Problem Formulation (Day 1–2)**
    –   Define decision variables, objectives, and constraints.
    –   Classify the problem (LP/NLP? Convex?).
2. **Numerical Solution (Day 3–4)**
    –   Use **Python/Pyomo** or **MATLAB** to solve:
        –   Maximize yield Y.
        –   Minimize cost (alternative objective).
    –   Generate sensitivity plots (e.g., yield vs. budget).

3. **Trade-Off Analysis (Day 5)**
   - Propose two Pareto-optimal solutions:
     - **Solution 1**: High yield, higher cost.
     - **Solution 2**: Low cost, moderate yield.
4. **Presentation (Day 6–7)**
   - Prepare a 10-minute pitch to "investors" (classmates) covering:
     - Methodology.
     - Results with visualizations.
     - Ethical considerations (e.g., sustainability vs. profit).

Deliverables

1 **Written Report** (3–5 pages):
   - Problem statement, formulation, code, results, and recommendations.
2 **Code Submission**:
   - Commented script (e.g., .py or .m file).
3 **Presentation Slides**:
   - Highlight key trade-offs and business implications.

4 Linear Programming (LP) Methods

Straight lines can solve the most twisted problems, if you know where to draw them.

4.1 Introduction to Linear Programming

A *linear program* (LP) is an optimization model with an objective and constraints in which all relationships are linear. The objective could be, for example, to maximize profit or to minimize costs. But other goals could also be set.

The key assumption in solving LPs is that the equations and inequalities are of course, linear. In addition, the principle of additivity, divisibility, and determinism must hold.

Additivity means that the total contribution of all variables in the objective function and constraints must be the sum of the individual contributions. For example, in a blending problem, the total cost of a mixture is the sum of costs of individual components, i.e., blending gasoline from two crudes (x_1 and x_2), the total cost is *$50x_1 + 30x_2$* (no cross-terms like x_1x_2). Many engineering problems show *non-additive effects* (e.g., chemical reactions that change costs based on interactions) require nonlinear models.

Divisibility means that the decision variables can take any fraction value and are not restricted to integers. Typically, flow rates, temperatures, and concentrations are naturally continuous. There are also many *cases* when variables exhibit a discrete nature, for example, the number of stages in a distillation column or the number of heat exchangers to install. In such a case, we need to look at methods for solving (mixed) integer programs.

Lastly, LPs must follow *determinism*, which means that all parameters (coefficients in the objective and constraints) are known with certainty. Also, here you can think of problems where the parameters are not exactly known, in such a case we could take recourse to stochastic programming (for probabilistic models) or robust optimization (for worst-case scenarios).

We can make full use of setting up and solving LPs in chemical engineering, think, for example, of blending problems (for crude oil or fuels), or resource allocation (equipment and raw materials), and production scheduling.

4.2 Formulating LP Problems

An LP model has three components: namely (1) the *decision variables* (e.g., x_1 = *flow rate of feedstock A*), (2) an *objective function* (e.g., maximize profit = $50x_1 + 30x_2$*), and (3) the *constraints* (e.g., *$2x_1 + x_2 \leq 100$* for resource limits).

https://doi.org/10.1515/9783111342283-004

Example 1: The Manufacturing Problem A manufacturing line can make two products, say product A and product B, from two resources, $R1$ and $R2$. For each unit of product A, we need 1 unit of $R1$ and 3 units of $R2$. For each unit of product B we need 1 unit of $R1$ and 2 units of $R2$. Product A sells for US\$6 per unit and product B sells for US\$5 per unit. We further have a limited amount of resources available. We have 5 units of R1 and 12 units of R2. The data is listed in Table 4.1.

Table 4.1: Data for the manufacturing problem.

	Resource 1	Resource 2	Profit
Product A	1	3	6
Product B	1	2	5
Available	5	12	

On the basis of this information we need to decide what we will consider decision variables, objective, and constraints.

Decisions: Let X be the number of units produced of product A and let Y be the number of units produced of product B.

Objective: From the provided sales price for the products, we can conclude that it is a good idea to maximize the profit, which can be computed as $Z = 6X + 5Y$

Constraints: We must make sure that the recipe is followed and that we are not using more resources than are available. This leads to the following constraints: $X + Y \leq 5$ and $3X + 2Y \leq 12$

We actually have another constraint. The so-called *non-negativity* constraint, which means that the decision variables cannot be negative: $X, Y \geq 0$
    The overall LP can now be formulated as

$$\max Z = 6X + 5Y$$

Subject to

$$X + Y \leq 5$$
$$3X + 2Y \leq 12$$
$$X, Y \geq 0$$

Let us have a look at a second example:

From the Expert: **Prof. Antonis Kokossis**
National Technical University of Athens, Greece

From Models to Machines:
Integrating PSE with AI for the Digital Chemical Enterprise

PSE technology is the principal reason the chemical process industry is also a data intensive industry. Indeed, data generated daily in chemical industries range from 1 to 20 TB in R&D, 50 to 500 TB in production, 1 to 20 TB in product portfolio, and 1 to 20 TB in supply chains. Such data structures set up a natural background for digitalization and AI that could promise further scope for transformational changes. As AI and ML hold a huge and disruptive potential to boost toward higher levels of efficiency and new value chains, chemical companies currently invest heavily in expertise that combines systems expertise with data capture and storage. Industries so far applied standard analyses and human intelligence to extract insights from this data. To that purpose, 95% of chemical companies have at least developed an AI strategy, while more than 50% are already in the pilot-program or early adoption stages, while two-thirds of them plan to invest more as high as 20% of digital budgets to AI.

While traditional AI focuses on analyzing or categorizing data by actively producing original outputs, its most exciting aspect is certainly generative AI, namely a type of artificial intelligence that creates new content (synthetic data) in response to various prompts. Such prompts naturally originate from models and PSE is a core instrument to explore an expanded space and new dimensions. AI and ML could indeed capitalize on the large and valuable legacy of process systems PSE models as available by means of process optimization and development models. Reversely, empowered by AI and ML, this legacy can be brought forward into a new generation of systems developments. Examples in exploring new dimensions could be numerous: expand the space of conventional energy integration to integrate renewables and non-conventional supplies; systems of systems integration across industrial domains; or development of digital twins for decentralized production, downscaling and control. Reversely, AI and ML can be used to produce scripts for software or produce models on the fly with language commands (e.g., by means of LLMs).

Chemical companies will need cross-functional expertise and people to work at the interface between process systems engineering and data analytics, able to drive transformation by training colleagues and helping to ensure the effective integration of systems challenges with AI. This new generation of experts could guide AI experts while knowledgeable about the potential to extract value from technologies. It is estimated that approximately 500–1,500 full-time employees with such skills will be required by a typical €10 billion turnover, 15,000 employee chemical company. These skills are difficult to recruit today, and they will become even scarcer as demand increases.

Reference
[1] A. Marousi, A. Kokossis, On the acceleration of global optimization algorithms by coupling cutting plane decomposition algorithms with machine learning and advanced data analysis, Computers & Chemical Engineering, 163, 107820, (2022)

Example 2: The Planning Problem A factory produces a single product. The demand for this product is estimated for a 4-month period: 1,000, 800, 1,200, and 900. The factory uses two production modes to make the product. It can produce in regular time with a capacity of 800 units per month, and it can produce in overtime with a capacity of 200 units per month. The costs of production are for regular time US$20 per unit and for overtime US$25 per unit. It is also possible to carry inventory to the next month. The inventory holding costs are US$3 per unit per month. We have to make sure that the demand is satisfied for each month and that at the end of the planning period, there are no more units in stock.

With this information, we have to decide the decision variables, the objective, and the constraints.

Decisions: Let X_j be the quantity of product produced in regular time mode in month j, let Y_j be the quantity of product produced in overtime mode in month j, and let I_j be the quantity of product kept in inventory in month j.

Objective: From the production costs and inventory holding costs, we can conclude that this is a cost-minimization problem. The objective function can be written as the sum of all products produced in regular time mode, overtime mode, and inventory holdings multiplied by their respective cost parameters: $Z = 20\sum_j X_j + 25\sum_j Y_j + 3\sum_j I_j$.

Constraints: We will have three types of constraints, the demand constraint for each month, the capacity of the production mode and of course the non-negativity.

The demand for month 1:

$$X_1 + Y_1 = 1,000 + I_1$$

The demand for month 2:

$$I_1 + X_2 + Y_2 = 800 + I_2$$

The demand for month 3:

$$I_2 + X_3 + Y_3 = 1,200 + I_3$$

And the demand for month 4 (making sure there is no inventory left):

$$I_3 + X_4 + Y_4 = 900$$

We must guarantee that the production capacity will not be exceeded $X_j \le 800 \forall j$ and $Y_j \le 200 \forall j$. The production cannot be negative, i.e., $X_j, Y_j, I_j \ge 0 \forall j$. You may have noticed the upside-down A symbol, which means: *for all*.

The result is an LP:

$$\min Z = 20\sum_j X_j + 25\sum_j Y_j + 3\sum_j I_j$$

Subject to:

$$X_1 + Y_1 = 1,000 + I_1$$
$$I_1 + X_2 + Y_2 = 800 + I_2$$
$$I_2 + X_3 + Y_3 = 1,200 + I_3$$

$$I_3 + X_4 + Y_4 = 900$$

$$X_j \leq 800 \forall j$$

$$Y_j \leq 200 \forall j$$

$$X_j,\ Y_j,\ I_j \geq 0 \forall j$$

This LP has 11 variables and 24 equations/constraints (1 objective, 4 demand constraints, 2 × 4 capacity constraints, and 11 non-negativity constraints).

We have now modeled the demand constraints as equalities, with help of the inventory variable. But we could alternatively have formulated them as inequalities without the inventory variable. We just track the difference between the left-hand side and right-hand side (RHS) of the demand constraint and add it to the next one.

The constraint for month 1 would become:

$$X_1 + Y_1 \geq 1,000$$

For month 2:

$$X_2 + Y_2 + (X_1 + Y_1 - 100) \geq 800$$

For month 3:

$$X_3 + Y_3 + (X_2 + Y_2 - 800 + X_1 + Y_1) \geq 1,200$$

For month 4:

$$X_4 + Y_4 + (X_3 + Y_3 - 1,200 + X_2 + Y_2 - 800 + X_1 + Y_1 - 1,000) = 900$$

This last constraint has to be an equality, as we do not want to have inventory left at month 4.

The inventory cost term in the objective now also will look different:

$$\min Z = 20 \sum_j X_j + 25 \sum_j Y_j + 3(X_1 + Y_1 - 1,000) + \cdots$$

$$\cdots + 3(X_1 + Y_1 + X_2 + Y_2 - 1,800) + \cdots$$

$$\cdots + 3(X_1 + Y_1 + X_2 + Y_2 + X_3 + Y_3 - 3,000)$$

This LP has 8 variables and 21 equations/constraints (1 objective, 4 demand constraints, 2 × 4 capacity constraints, and 8 non-negativity constraints).

Actually we could have written a third version of the model, where we consider the variable X_{ijk} which is the amount of product produced in month i, for month j, with production mode k. By choosing to use this variable with three indices, there are of course consequences for the constraints and objective function. Let us look again at the demand constraints.

For month 1:

$$X_{111} + X_{112} = 1,000$$

For month 2:

$$X_{121} + X_{122} + X_{221} + X_{222} = 800$$

For month 3:

$$X_{131} + X_{132} + X_{231} + X_{232} + X_{331} + X_{332} = 1,200$$

For month 4:

$$X_{141} + X_{142} + X_{241} + X_{242} + X_{341} + X_{342} + X_{441} + X_{442} = 900$$

Also the capacity constraints have now to be written for each month. For regular time they become

$$X_{111} + X_{121} + X_{131} + X_{141} \leq 800$$
$$X_{221} + X_{231} + X_{241} \leq 800$$
$$X_{331} + X_{341} \leq 800$$
$$X_{441} \leq 800$$

And for over time:

$$X_{112} + X_{122} + X_{132} + X_{142} \leq 800$$
$$X_{222} + X_{232} + X_{242} \leq 800$$
$$X_{332} + X_{342} \leq 800$$
$$X_{442} \leq 800$$

Also the objective function now a bit different now:

$$Z = 20(X_{111} + X_{121} + X_{131} + \cdots$$
$$\cdots + X_{221} + X_{231} + X_{241} + \cdots$$
$$\cdots + X_{331} + X_{341} + \cdots$$
$$\cdots + X_{441}) + \cdots$$
$$\cdots + 25(X_{112} + X_{122} + X_{132} + \cdots$$
$$\cdots + X_{222} + X_{232} + X_{242} + \cdots$$
$$\cdots + X_{332} + X_{342} + \cdots$$
$$+ X_{-}442) + \cdots$$
$$\cdots + 3(X_{121} + X_{122} + X_{231} + X_{232} + X_{341} + X_{342}) + \cdots$$
$$\cdots + 6(X_{131} + X_{132} + X_{241} + X_{242}) + \cdots$$
$$\cdots + 9(X_{141} + X_{142})$$

It is noted that the constraints and equation were written explicitly, showing all variables. It is possible to write the model more compact by using summations.

What can be concluded from these three examples?

That LP models can be formulated in different ways.

But what is the best way? Two rules to help you out a bit are:

1. Try to formulate your LP with the smallest number of variables and constraints as needed;
2. Generally, it is better to use inequalities than equalities.

4.3 Graphical Solution Method

So once an LP is formulated, we want to look for a solution to the problem. We might use a graphical method. Let us consider the first example from the previous section, the manufacturing problem:

$$\max 6X_1 + 5X_2$$

subject to:

$$X_1 + X_2 \le 5$$

$$3X_1 + 2X_2 \le 12$$

$$X_1, X_2 \ge 0$$

We could draw the constraints in a plane, plotting X_1 versus X_2. For the first constraint, we could find two points, first by fixing $X_1 = 0$, that means that $X_1 + X_2 = 5, X_2 = 5$. So that leads to a point $(0,5)$, and the other way around, fixing $X_2 = 0$, meaning that $X_1 + X_2 = 5, X_1 = 5$. This gives the point $(5,0)$. We can connect the two points via a straight line in our plane.

We can do the same thing for the other constraint. first by fixing $X_1 = 0$, that means that $3X_1 + 2X_2 = 12, X_2 = 4$. So that leads to a point $(0,4)$, and the other way around, fixing $X_2 = 0$, meaning that $3X_1 + 2X_2 = 12, X_1 = 6$. This gives the point $(6,0)$. We can connect the two points via a straight line in our plane. Figure 4.1. shows the X_1-X_2 plain with the two constraints.

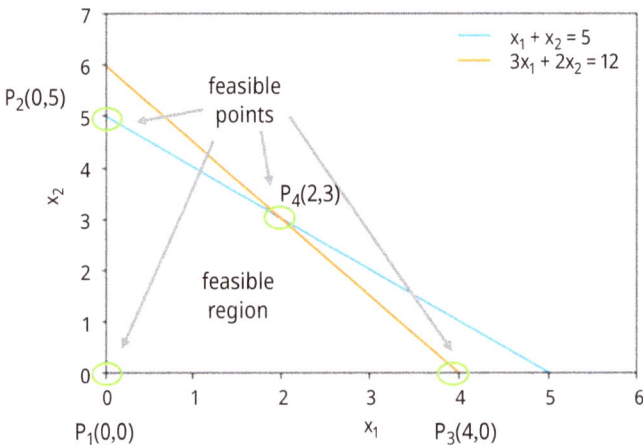

Figure 4.1: X_1-X_2 plane showing the two constraints.

You must not forget that X_1 and X_2 have to be positive. In other words, we only look at values in the first quadrant, i.e., above the X-axis and right of the Y-axis. As the constraints are inequalities, we must consider only values below these lines. The *feasible region* is the area inside the plane that satisfies all constraints. And all points in this region are called *feasible points*. For example, the point (1,1) is a feasible point. For this point, the objective function value is $Z = 6*1 + 5*1 = 11$. But are there better points? Yes! If you move up and/or to the right, you can improve the solution till you reach the boundary. For any point inside the feasible region, there is always a point on the boundaries that will dominate the objective function. So we should focus on the boundaries.

The corner points are the most interesting. These so-called *vertices* (singular *vertex*) will dominate the objective function values.

Point	Objective value
0,0	$Z = 0$
4,0	$Z = 24$
0,5	$Z = 25$
2,3	$Z = 27$

You can draw the objective function in the same X_1–X_2 plane for different values of Z. That is demonstrated in Figure 4.2. The corner point (2,3) is the point where the objective function touches last when leaving the feasible region. The point (2,3) is the best possible solution and is called the *optimum*.

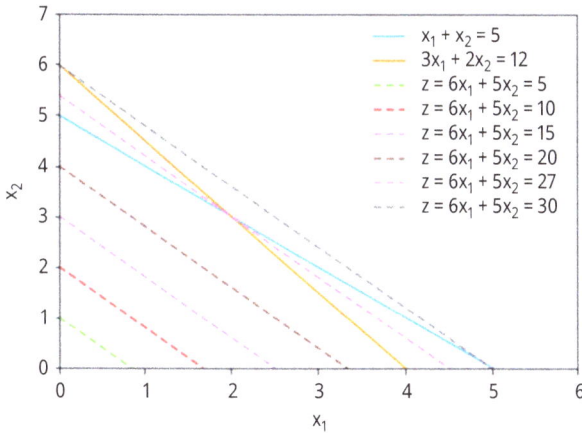

Figure 4.2: The X_1–X_2 plane with objective function for different values of Z.

So, in summary: the graphical method consists of three steps:
1. Plot constraints in a graph;
2. Identify feasible region and corner points;
3. Identify the best corner point.

There is however a main limitation with the graphical method. It only works with two variables. Many engineering problems are significantly larger. So we need another solution approach.

4.4 The Algebraic Method

What we will first do, for our manufacturing example we will write the inequality constraints into equalities, by introducing two additional variables:

$$\max 6X_1 + 5X_2$$

subject to:

$$X_1 + X_2 + X_3 = 5$$

$$3X_1 + 2X_2 + X_4 = 12$$

$$X_1, X_2, X_3, X_4 \geq 0$$

X_3 and X_4 are called *slack variables*. So, how are they defined? And, do they contribute to the objective function? Suppose X_1 and X_2 are the amounts of products A and B that a factory produces. What are X_3 and X_4, then? Suppose we make 4 units of product A and 0 of product B, then following the first constraint:

$$X_1 + X_2 + X_3 = 5$$

It means that $X_3 = 1$, which means that not all resources are utilized and left available. In general, we do not add the slack variables to the objective function.

In our LP we have two equations (the constraints) and four unknowns. So we should fix two variables and solve the other two. But which ones to fix and to which values?

If we have to select two variables from a set of four we can make six different pairs, according to

$$nC = \frac{n!}{(n-r)!r!} = 4C_2 = \frac{4!}{(4-2)!2!} = 6$$

Let us fix the pairs of variables to zero and solve for the other ones, for each of the six combinations:

Combination	Fixed		Solved		Constraint satisfied	Objective value
1	$X_1 = 0$	$X_2 = 0$	$X_3 = 5$	$X_4 = 12$	Yes	$Z = 0$
2	$X_1 = 0$	$X_3 = 0$	$X_2 = 5$	$X_4 = 2$	Yes	$Z = 25$
3	$X_1 = 0$	$X_4 = 0$	$X_2 = 0$	$X_3 = -1$	No	
4	$X_2 = 0$	$X_4 = 0$	$X_1 = 4$	$X_3 = 1$	Yes	$Z = 24$
5	$X_2 = 0$	$X_3 = 0$	$X_1 = 0$	$X_4 = -1$	No	
6	$X_3 = 0$	$X_4 = 0$	$X_1 = 2$	$X_2 = 3$	Yes	$Z = 27$

The combinations 1, 2, and 4 lead to feasible solutions. The combinations 3 and 5 lead to infeasible solutions (as the variables have to be positive). Combination 6 leads to the optimal solution.

Maybe you have noticed, there is a relationship between the algebraic method and the graphical method. Take a look at Figure 4.3. The combinations are actually the corner points in the X_1–X_2 plane. The infeasible solutions are the ones that only satisfy one constraint.

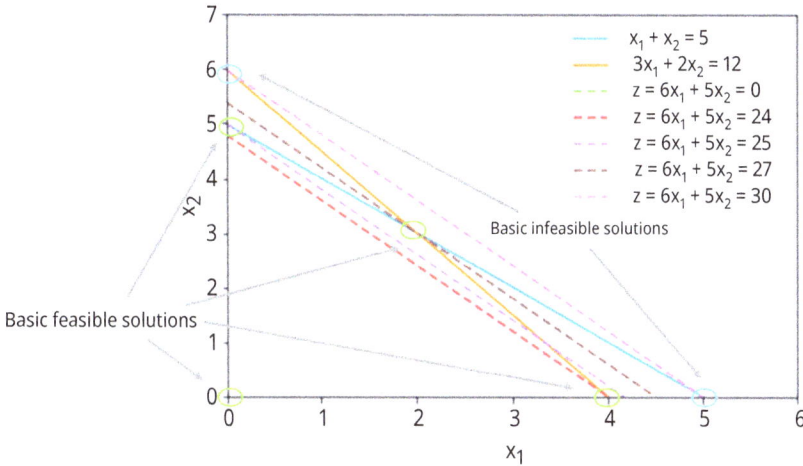

Figure 4.3: The basic feasible and basic infeasible solutions.

It turns out that for any feasible X_1 and X_2 you should fix X_3 and X_4 to a non-negative value. For any corner point you fix two variables (of the six combinations) to zero and solve for the others.

The six solutions are called *basic solutions*. All other solutions are called *non-basic solutions*. We found that four solutions were feasible (called *basic feasible solutions*) and that two solutions were infeasible (which are called *basic infeasible*). The variables that were fixed to zero are called *non-basic variables* and the variables that were solved are called *basic variables*.

In summary, the procedure to solve a LP algebraically is as follows:
1. Add slack variables (four variables and two equations);
2. Fix any two arbitrary variables and solve the remaining two. Fixed variables can be chosen in $4C_2$ ways;
3. In each combination we can fix variables to any values, but we choose to fix at zero, leading to six solutions;
4. Fixing to zero leads to basic solutions;
5. Identify the optimum.

Although the algebraic method can be used to solve any number of variables and equations, there are a number of disadvantages:
1. nC_r can be a very large number, i.e., many solutions to be computed;
2. Also infeasible solutions are calculated;
3. The solution does not become better after each calculation step.

The *Simplex method* considers these drawbacks.

Figure 4.4: PhD candidate Hung and the Shadow pricing in linear programs.

In figure 4.4. you can see how PhD student Hung learns about setting up and solving linear programs.

4.5 Simplex Method

The Simplex method is an algorithmic procedure for solving LP problems (with more than two variables). It was developed in the 1960s by *George Dantzig*, with the aim of optimizing logistic operations of the military. The Simplex method will exclude the evaluation of infeasible solutions, it will also find a better solution at each iteration and it will have a criterion that stops the iteration as soon as the optimum is found.

We will consider the previous example with slack variables here:

$$\max 6X_1 + 5X_2$$

subject to:

$$X_1 + X_2 + X_3 = 5$$

$$3X_1 + 2X_2 + X_4 = 12$$

$$X_1, X_2, X_3, X_4 \geq 0$$

We do not want to evaluate infeasible solutions; in other words, solutions were $X_j < 0$. We can start with finding a basic feasible solution by fixing $X_1 = X_2 = 0$. We find that $X_3 = 5$ and $X_4 = 12$. In the Simplex method, constraints cannot have negative values for the RHS, if so, multiply with −1.

Iteration no. 1:
We express the constraints in terms of the basic variables:

$$X_3 = 5 - X_1 - X_2 \tag{4.1}$$

$$X_4 = 12 - 3X_1 - 2X_2 \tag{4.2}$$

If we look at the objective function $Z = 6X_1 + 5X_2$ we can increase Z by increase X_1 and X_2. In the Simplex method, we only increase one variable at the time. X_1 is the most logical because it increases Z the most.

If we increase X_1, it will reduce X_3 and X_4. So we can only increase X_1 to a value that X_3 and X_4 become zero (otherwise we would create an infeasible problem).

If we look at constraint (4.1) we can only increase $X_1 \rightarrow 5$. If we look at constraint (4.2) we can only increase $X_1 \rightarrow 4$. In other words, we cannot increase X_1 to a value above 4. Constraint (4.2) is the limiting constraint which we will write in terms of X_1:

$$3X_1 = 12 - 2X_2 - X_4$$

Rewrite:

$$X_1 = 4 - \frac{2}{3}X_2 - \frac{1}{3}X_4 \tag{4.3}$$

We will keep constraint (4.1) and substitute constraint (4.3) in it:

$$X_3 = 5 - \left(4 - \frac{2}{3}X_2 - \frac{1}{3}X_4 \right) - X_2$$

Rewrite:

$$X_3 = 1 - \frac{1}{3}X_2 + \frac{1}{3}X_4 \tag{4.4}$$

We now substitute constraints (4.3) and (4.4) to get the objective function:

$$Z = 6 \left(4 - \frac{2}{3}X_2 - \frac{1}{3}X_4 \right) + 5X_2$$

Rewrite:

$$Z = 24 - X_2 - 2X_4 \tag{4.5}$$

The variables that appear in the objective function are the non-basic variables (X_2 and X_4) and the other ones have become the basic variables (X_1 and X_3).

We fix the non-basic variables to zero and solve for the others. We can solve for the others by filling $X_2 = X_4 = 0$ into constraints (4.3) and (4.4). We find: $X_1 = 4$ and $X_3 = 1$. From the objective function (eq. (4.5)) we then read that the objective value is $Z = 24$. This solution is a basic feasible solution.

Iteration no. 2:
If we want to increase Z (from eq. (4.5)) we should increase X_2 or decrease X_4. X_4 cannot be negative and already has a value of 0, so we should focus on X_2.

From constraint (4.3), we see that X_2 can be increased up to $X_2 \rightarrow 6$ before X_1 will be negative. And from constraint (4.4) we see that X_2 can be increased up to $X_2 \rightarrow 3$ before X_3 becomes negative. So we will use constraint (4.4) as limiting constraint. We rewrite constraint (4.4) in terms of X_2:

$$\frac{1}{3}X_2 = 1 = X_3 + \frac{1}{3}X_4$$

Rewrite:

$$X_2 = 3 - 3X_3 + X_4 \tag{4.6}$$

We modify constraint (4.3) by substituting constraint (4.6) into it:

$$X_1 = 4 = \frac{2}{3}(3 - 3X_3 + X_4) - \frac{1}{3}X_4$$

Rewrite:

$$X_1 = 2 - 2X_3 - X_4 \tag{4.7}$$

We substitute (4.6) and (4.7) in the objective function (4.5):

$$Z = 24 + (3 - 3X_3 + X_4) - 2X_4$$

Rewrite:

$$Z = 27 - 3X_3 - X_4 \tag{4.8}$$

X_3 and X_4 are the non-basic variables, and X_1 and X_2 are the basic variables. We fix the non-basic variables to zero, and from constraint (4.6), that $X_2 = 3$ and from constraint (4.7) that $X_1 = 2$. From the objective function (4.8), we read that $Z = 27$.

Can we further increase Z? If we check the objective function (4.8), it seems that Z can only be improved if we make X_3 and/or X_4 smaller. But they are already zero. So it is not possible to increase the objective further. We have found the optimum, and the iteration can be stopped.

The Simplex method is a smart way to identify which variables affect the objective and then rearrange and substitute the constraints in such a way that we improve the objective function. During each calculation step, the variables that affect the objective change. At some point, the variables turn zero, and further increasing the objective is no longer possible.

We have now written the equations separately, but for larger problems it is worthwhile to use a *tabular form*. The initial *Simplex table* (or *Simplex tableau*) looks as follows:

	X_1 (6)	X_2 (5)	X_3 (0)	X_4 (0)	RHS
X_3 (0)	1	1	1	0	5
X_4 (0)	3	2	0	1	12
$C_j - Z_j$	6	5	0	0	0

In the top row, we write all variables and which value they add to the objective function. In the first column, we write the basic variables. We also write the values that the basic variables have in the objective function. In the core of the table we list the coefficients of all the constraints. In the last row, we will calculate the dot product ($C_j - Z_j$). We use the entries of the column in question and the values of the basic variables.

For example, for the first column:

$$C_j - Z_j = 6 - (0*1 + 0*3) = 6$$

The second column:

$$C_j - Z_j = 5 - (0*1 + 0*2) = 5$$

For the third column:

$$C_j - Z_j = 0 - (0*1 + 0*0) = 0$$

And for the fourth column:

$$C_j - Z_j = 0 - (0*0 + 0*1) = 0$$

We now look in the Simplex table to the $C_j - Z_j$ that has the largest value, i.e., 6 in the first column. This element enters the basis (we often denote that with a little up arrow). X_1 will now be our pivot column.

	X_1 (6)	X_2 (5)	X_3 (0)	X_4 (0)	RHS	θ
X_3 (0)	1	1	1	0	5	5
X_4 (0)	3	2	0	1	12	4→
$C_j - Z_j$	6↑	5	0	0	0	

We now divide the RHS by the values from the pivot column and add these coefficients to the table as an extra column θ. From these coefficients we select the smallest number and use it as the pivot row. That is the second row in this case, with X_4. X_4 will leave the basis (this is denoted with a right arrow). So X_1 enters and X_4 leaves. It is possible that in subsequent iterations variables that earlier left, can re-enter the table.

The shaded part in the table is an identity matrix. We will now perform row operations to create a new identity matrix (with the new basic variables).

Row	No. 1	X_1 (6)	X_2 (5)	X_3 (0)	X_4 (0)	RHS	θ
1	X_3 (0)	1	1	1	0	5	5
2	X_4 (0)	3	2	0	1	12	4→
	$C_j - Z_j$	6↑	5	0	0	0	
	No. 2						
3	X_3 (0)	0	1/3	1	−1/3	1	
4	X_1 (6)	1	2/3	0	1/3	4	

Divide row 2 by the pivot element (orange) and store entries in row 4. Then subtract row 4 from row 1 and store in row 3. Now I have made a new identity matrix with the new basic variables. Now we will repeat the procedure with the new Simplex table.

1. Calculate $C_j - Z_j$
2. Select column with the largest $C_j - Z_j$ (i.e., X_2, which enters the basis)
3. Calculate θ and divide the RHS by the pivot column
4. Select row with smallest θ (i.e., X_3, which leaves the basis)
5. Now create a new identity matrix with the new basic variables using row operations
   a. Multiply row 3 with 3 to create row 5
   b. Subtract row 3 two times from row 4 to create row 6
6. Repeat from step 1.

The iteration stops when there are no longer positive values for $C_j - Z_j$, then the optimum is found.

Row	No. 1	X_1 (6)	X_2 (5)	X_3 (0)	X_4 (0)	RHS	θ
1	X_3 (0)	1	1	1	0	5	5
2	X_4 (0)	3	2	0	1	12	4→
	$C_j - Z_j$	6↑	5	0	0	0	
	No. 2						
3	X_3 (0)	0	1/3	1	−1/3	1	3→
4	X_1 (6)	1	2/3	0	1/3	4	6
	$C_j - Z_j$	0	1↑	0	−2	24	
	No. 3						
5	X_2 (5)	0	1	3	−1	3	
6	X_1 (6)	1	0	−2	1	2	
	$C_j - Z_j$	0	0	−3	−1	27	

In summary the Simplex method comprises four steps:
1. Convert inequalities to equations using slack variables.
2. Construct the Simplex tableau.
3. Pivot operations to improve the objective function.
4. Terminate when no further improvements (optimality).

4.6 Sensitivity Analysis

Sensitivity analysis can be used to understand how changes in LP parameters (cost coefficients and constraint limits) impact the optimal solution. The key questions are: (1) How much can a cost coefficient change before the optimal solution changes? (2) What is the value of increasing a resource limit (e.g., raw materials and capacity)? (3) How robust is the solution to uncertainties in input data?

4.6.1 Shadow Prices (Dual Values)

The *shadow price* of a constraint is the change in the objective value (Z) per unit increase in the RHS of that constraint, assuming all other conditions remain constant.

The shadow price quantifies the marginal value of resources (e.g., extra reactor capacity and feedstock supply) and helps prioritize investments (e.g., is buying more raw material worth the cost?).

Example: Consider the following example. In a factory, we produce two chemicals. We have limitations in the raw material availability and in the labor hours. The goal is to maximize the profit by producing X and Y amounts of chemicals. The problem is formulated as an LP.

The profit is given as

$$\max Z = 80x + 60y$$

In €/day.

We have raw material limitations:

$$2x + y \le 40$$

which means production cannot exceed 40 tons/day.

And we have limits in labor:

$$x + y \le 25$$

Which means that labor cannot exceed 25 h/day.

We can solve the above LP with the graphical method or with the Simplex method and find that the optimal solution is: $x = 15$, $y = 10$, and $Z = 1,800$.

The shadow price is the rate of change of the objective function with respect to a small increase in the RHS of a constraint.

If we increase the RHS of the raw materials constraint to 41 ton/day and solve the LP, we will find new values for $x = 16$, $y = 9$, and $Z = 1,820$. This means that if we increase the raw materials capacity by 1 ton, the profit increases by €20:

$$\frac{\Delta Z}{\Delta b} = \frac{1,820 - 1,800}{41 - 40} = 20 \text{ €/ton}$$

If we increase the RHS of the labor constraint to 26 h/day and solve the LP, we will find $x = 14$, $y = 12$, and $Z = 1,840$. If we increase the labor limit by 1 h, the profit increases by €40:

$$\frac{\Delta Z}{\Delta b} = \frac{a1,840 - 1,800}{26 - 25} = 40 \text{ €/day}$$

We must understand that shadow prices are valid only within specific ranges of the RHS. Beyond these, the optimal basis (active constraints) changes.

For the raw materials constraints, the RHS can only be increased by 5 tons (up to 45). If the RHS goes above 45 the labor constraint becomes the only binding constraint.

In the same way, for the labor constraint, the RHS can only increase with 5 h (up to 30) or decrease with 5 h (down to 20) before the raw materials constraint becomes the only binding constraints.

4.6.2 Reduced Costs

Let us consider the following example:

A petrochemical plant blends two feedstocks (A and B) to maximize profit. The profit for A is \$50/ton and for B is \$30/ton. The objective can be written as

$$\max Z = 50x_A + 30x_B$$

where x_A and x_B are the amounts of A and B being blended. For this problem, there is a capacity limit:

$$x_A + x_B \leq 100 \text{ ton/day}$$

There is also a quality requirement constraint that reads:

$$x_A \geq 0.4(x_A + x_B)$$

which means that feedstock A must be more or equal than 40% of the blend. The optimal solution can be found via the graphical method or via the Simplex method: $x_A = 40$ tons, $x_B = 60$ tons, and $Z = \$3,800$.

The *reduced costs* of a non-basic variable (a variable not in the optimal solution) indicate how much it objective coefficient must improve (increase for maximization and decrease for minimization) for the variable to enter the basis.

Basic variables (variables in the optimal solution always have reduced costs of zero. Non-basic variables (variables that are zero in the optimal solution) have reduced cost that are zero or negative (maximization) or zero or positive (minimization).

Below you will find the final Simplex Tableau:

Basis	x_A	x_B	s_1	s_2	RHS
x_A	1	0	1	−1	40
x_B	0	1	−0.4	1.4	60
Z	0	0	20	10	3,800

The coefficients of x_A and x_B in the Z-row are zero (both basic variables). If we had a third variable x_C (non-basic) the reduced cost would appear in the Z-row.

Suppose the plant considers a third feedstock C with a profit of \$20/ton. $x_A = 40$ tons, $x_B = 60$ tons, $x_C = 0$, and $Z = \$3,800$.

For example, the final Simplex Tableau would look like:

Basis	x_A	x_B	x_C	s_1	s_2	RHS
x_A	1	0	1	1	−1	40
x_B	0	1	−0.4	−0.4	1.4	60
Z	0	0	−10	30	50	3,800

It turns out that the reduced cost for x_C is −10 (the entry in the Z-row). This means that the profit of x_C must increase by \$10/ton (from \$20 to \$30/ton) to enter the basis.

The shadow prices can also be directly read from the Simplex Tableau. The shadow prices for the capacity constraint would be \$30/ton and for the quality constraint \$50/ton (read in the Z-row the s1 and s2 columns).

The reduced costs can be directly calculated from the profit and shadow prices and the coefficients in the final Simplex tableau, for a minimization problem:

$$\text{Reduced costs} = c_j - \pi^T A_j$$

And for a maximization problem:

$$\text{Reduced costs} = \pi^T A_j - c_j$$

So the reduced costs for x_C can be calculated as

$$30*1 + 50*(-0.4) - 20 = -10$$

Analyzing the reduced costs can be used to quantify the opportunity cost of unused resources and it guides cost-benefit decisions.

4.7 Linear Programming in Excel

The Excel solver function is a nice and useful tool within Excel. If you have not installed the solver add-in, you can consult Excel Help to make sure the solver is available. The solver add-in allows you to solve optimization problems. We will illustrate how the Excel solver works with a simple LP example.

Suppose you want to solve the following LP:

$$\max P = 3x_1 + 2x_2 - x_3$$

Subject to:

$$x_1 + 3x_2 + x_3 \leq 9$$

$$2x_1 + 3x_2 - x_3 \geq 2$$

$$3x_1 - 2x_2 + x_3 \geq 5$$

$$x_1, x_2, x_3 \geq 0$$

You could start with something like figure below. Where we define x_1, x_2, and x_3 and set them to a value of zero. We then calculate the value of the objective function: So cell $B5$ is calculated as: $= 3*B1 + 2*B2 - B3$. Figure 4.5 shows how to start setting up a linear program in Excel, start with variables and objective.

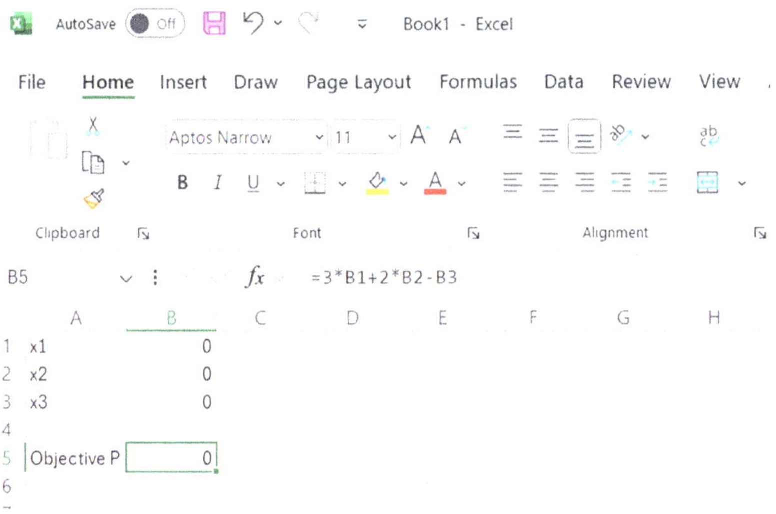

Figure 4.5: Excel sheet with variables and objective.

In a similar way we can define the constraints. We calculate the left-hand side value of the first constraint in cell $A9$ as: $= B1 + 3^*B2 + B3$, we list the sign and also the RHS value of the constraint. In figure 4.6. we now add the constraints.

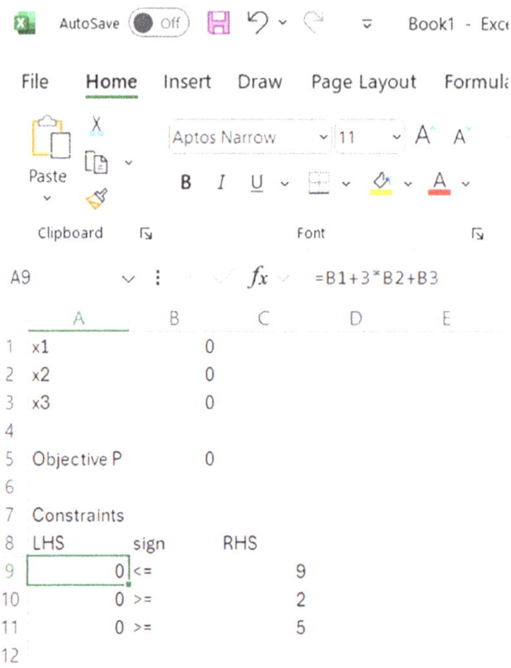

Figure 4.6: Excel sheet with now also the constraints added.

Now we open the solver add-in (in the top right corner). Figure 4.7 shows the tab with the solver on the right side.

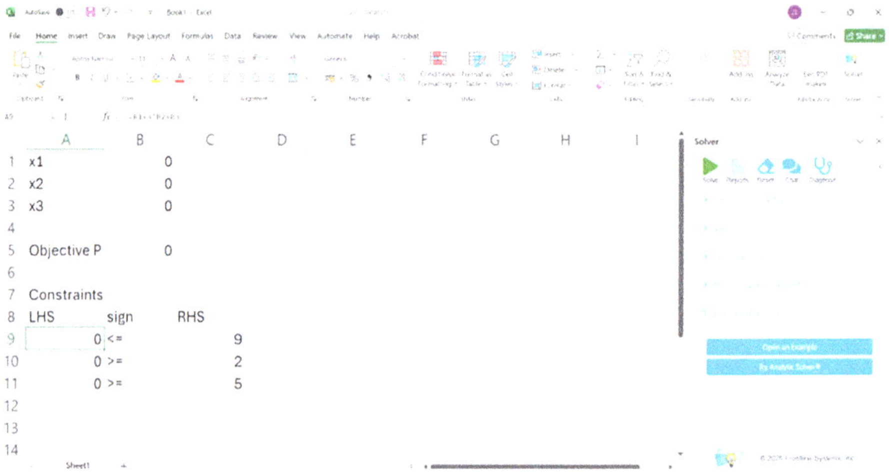

Figure 4.7: Excel sheet now with the solver plug-in open (right).

We have to provide information regarding, the objective, variables, and constraints. When we click on the Objective tab, we are asked to select the cell that we want to maximize (or minimize). In our case this is cell $B5$.

Under the tab variables, we have to list the variables that we want to use, so cells $B1$, $B2$, and $B3$. We can also tell Excel that these variables are positive (non-negative). Figure 4.8. shows how to select variables and objective in the solve plug-in.

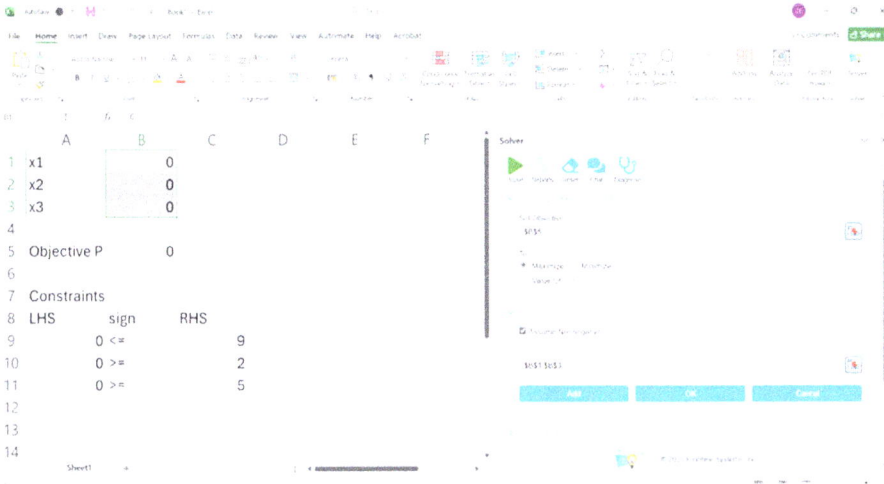

Figure 4.8: Excel solver plug-in, selecting the variables and objective.

Lastly, we have to feed the constraints to the solver, providing the LHS, the sign and the RHS (See figure 4.9):

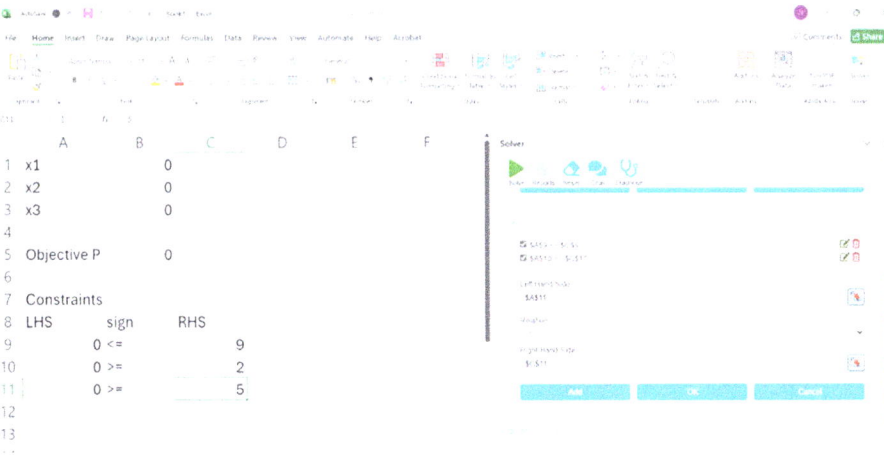

Figure 4.9: Excel sheet with solver plug-in: selecting the constraints.

You can proceed to solve the LP by clicking on the solve button. Excel, will automatically feed the solution back into the Excel sheet (See figure 4.10).

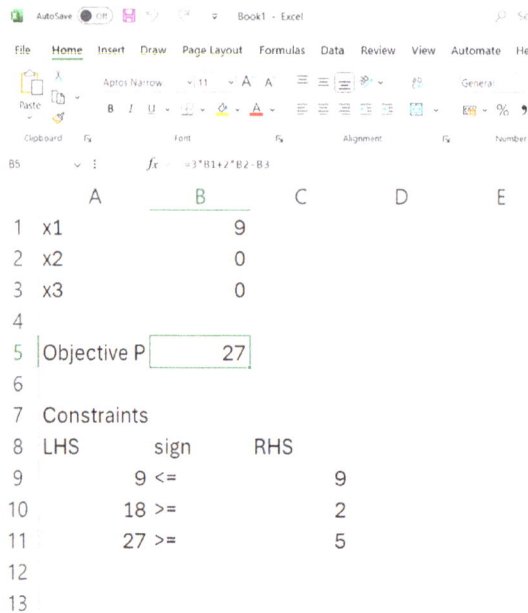

	A	B	C	D	E
1	x1	9			
2	x2	0			
3	x3	0			
4					
5	Objective P	27			
6					
7	Constraints				
8	LHS	sign	RHS		
9	9 <=		9		
10	18 >=		2		
11	27 >=		5		
12					
13					

Figure 4.10: Excel sheet with the optimized variables and objective.

The first constraint is limiting the problem. Excel found that $x_1 = 9$ and that x_2 and x_3 should be zero, leading to an objective of $P = 27$.

4.8 Exercises

Exercise 1: Biobased Fuel Production ★
In a biorefinery two biofuels (bio-ethanol and bio-diesel) can be produced via two different biobased raw materials (weathstraw and algae). To produce 1 ton of bioethanol 15 tons of weathstraw and 1 ton of algae are needed. To produce 1 ton of biodiesel 5 tons of weathstraw and 15 tons of algae are needed. The bioethanol sells for €800/ton and the biodiesel sells for €750/ton. The biorefinery has in total 75 tons of weathstraw and 100 tons of algae in stock. Formulate a LP for this problem.

Exercise 2: Renewable Energy Production ★★
A small energy company produces renewable energy via a wind mill farm and a solar farm. The company has costumers that have a weekly demand for energy. The weekly demand for the coming five weeks is given: 5,000 MWh, 4,200 MWh, 5,500 MWh, 6,000

MWh, and 4000 MWh. The wind mill farm and solar farm have also weekly maximum capacities of 2,500 and 3,500 MWh, respectively. The costs for producing renewable energy are for the wind mill farm €150/MWh and for the solar farm €175/MWh. The energy company has access to power storage facilities (e.g., via battery storage). This allows the company to store excess production with time. However there are costs associated with the energy storage of €100/MWh/week. Formulate this problem as an LP.

Exercise 3: Graphical Solution to the Biobased Fuel Production Linear Program ★★

We found from the previous exercise the following LP:

$$\max 800X + 750Y$$

subject to

$$15X + Y \leq 75$$

$$5X + 15Y \leq 100$$

$$X, Y \geq 0$$

where X and Y are the quantities of bio-ethanol and bio-diesel that can be produced from two biomass resources.

We now want to find the optimum values for X and Y that maximize the objective. Use the graphical method to identify the optimum. Plot the constraints, show the feasible region, and identify the corner points. Also plot the objective function in this window.

Exercise 4. Graphical Solution to the Biobased Fuel Production Linear Program ★★

Also use the algebraic method to solve the problem of Exercise 3. Show each step in your calculations as discussed in the lecture.

Exercise 5: Setting Up the Linear Program for a Hydrogen Production Plant ★★★

A hydrogen production facility produces two grades of hydrogen: grade 1 (very pure) and grade 2 (pure). The hydrogen sells for 8 k€/ton (very pure) and 6 k€/ton (pure) at the market. The facility uses an electrolysis plant to produce the hydrogen. To produce 1 ton of pure hydrogen 1 ton of acid, 1 ton of water, and 2 kW of electricity are required. To produce 1 ton of very pure hydrogen 1 ton of acid, 5 tons of water, and 3 kW of electricity are required. The facility has 10 tons of acid and 35 tons of water in stock. At this stage the available renewable electricity is 25 kW. Formulate this problem as a profit maximization problem (with profit in k€).

Exercise 6: Algorithmic Solution to the Hydrogen Production Problem ★★★
Use the tabulated simplex method to find the solution to Exercise 1. Show each step in your calculations as discussed in the lecture.

Exercise 7: Methanol Production ★★★
A methanol plant produces methanol and can sell it for $100/ton. There are two main constraints in the model: the supply of syngas and the reactor capacity. This leads to the following LP:

$$\max Z = 100x$$

Subject to:

$$3x \leq 90 \text{ (syngas supply)}$$

$$x \leq 25 \text{ (reactor capacity)}$$

The current optimum is $x = 25$ tons with $Z = \$2,500$.

Calculate the shadow prices for both constraints. If the syngas supply increases to 93 ton, what would be the new profit?

Exercise 8: Reduced Costs ★★★★
We are considering the following LP:

$$\max Z = 10x_1 + 12x_2$$

Subject to

$$x_1 + x_2 \leq 50$$

$$2x_1 + x_2 \leq 80$$

$$x_1, \ x_2 \geq 0$$

The optimal solution is $x_1 = 30$ and $x_2 = 20$, with $Z = \$540$. If a third variable x_3 (profit = 8$/unit) is introduced, what are the reduced costs? How much must x_3's profit increase to be produced?

4.9 Takeaway

In this chapter we have introduced the LP problem. Many engineering problems can be formulated as LPs. When setting up an LP we have to consider what is the objective, what are the decisions to make, and what are the constraints that we want to include. Generally it is good to keep the number of decisions and constraints as small as possible and it is better to have inequalities above equalities. Once the problem is set up, a solution can be found. For LP, the optimum always lies at the cross point of the constraints. We discussed a graphic method that is very insightful, but limited in problem size (two decision variables). We then discussed the algebraic method that can solve larger problems. The drawback of the algebraic method is that it also computes infeasible solutions and that its solutions do not necessarily improve at each calculation step. The Simplex method takes care of these shortcomings and leads efficiently to the optimum. When the solution is obtained a sensitivity analysis can be performed. The shadow prices (marginal values) can be used to study the effect of the RHS of the constraints on the objective function. And the reduced costs can be used to study the effect of the objective function coefficients on the objective function value. Such analysis can help in further improving your process. It turns out that the solver add-in from Excel is an easy tool to setup and solve LPs.

Further Reading

T. Edgar, D. Himmelblau, T. Lasdon, Optimization of Chemical Processes (2001), McGraw-Hill.

I. Grossmann, Advanced Optimization for Process Systems Engineering (2021), Cambridge University Press.

E. Kreyszig, Advanced Engineering Mathematics, 10th edition (2011), Wiley, Chapters 20 and 21.

5 Nonlinear Programming (NLP)

Because real-world processes are rarely linear.

5.1 Introduction to NLP

In the previous chapter, we got introduced to linear programming (LP). LP cannot handle curved feasibility regions or exponential objectives. It turns out that chemical engineering systems often involve **nonlinear relationships**: You can think of reaction kinetics (Arrhenius equation), thermodynamics (fugacity and activity coefficients), heat/mass transfer (log-mean temperature difference), or economic considerations (economies of scale). In this chapter, we will explore nonlinear programming (NLP) and its solution procedures. Key differences between LP and NLP are listed in Table 5.1.

Table 5.1: Key differences from LP.

Feature	LP	NLP
Objective	Linear	Nonlinear (e.g., quadratic and exponential)
Constraints	Linear inequalities	Nonlinear equations
Solution	Global optimum guaranteed	Local/global optima possible
Solvers	Simplex method	Gradient-based/heuristic methods

5.2 Formulating NLP Problems

Just as with other optimization problems, an NLP has three key components: 1) the decision variables (e.g., temperature or flow rate); 2) an objective function (e.g., maximize the yield, or minimize the constraints); and (3) constraints (e.g., safety limits or material balances).

Example: An example of an NLP is given below. For a tank reactor, the goal is to maximize the production rate of B (for a reaction A goes into B), which follows nonlinear kinetics. The concentration of A in the feed, as well as the temperature, are constrained. An NLP could look as follows:

$$\max r_B = k_0 e^{-\frac{E}{RT}} C_A^2$$

subject to:

$$C_A \leq C_{A,FEED}$$

https://doi.org/10.1515/9783111342283-005

$$T_{\min} \leq T \leq T_{\max}$$

$$Q_{\text{generated}} = -\Delta H_r r_B \leq Q_{\max}$$

with the following parameter values:

Parameter	Value	Units	Description
k_0	5.0	L/(mol · s)	Pre-exponential factor
E	5,000	J/mol	Activation energy
R	8.314	J/(mol · K)	Gas constant
$C_{A,\text{feed}}$	2.0	mol/L	Feed concentration
$T_{\min}$	300	K	Minimum safe temperature
$T_{\max}$	400	K	Maximum safe temperature
ΔH_r	−800	J/mol	Heat of reaction (exothermic)
$Q_{\max}$	2	kW	Cooling capacity

With the following Python code, this example can be solved, using the `minimize` function:

```python
import numpy as np
from scipy.optimize import minimize
# Realistic parameters
k0 = 0.05            # [L/(mol·s)]
Ea = 50000           # [J/mol]
R = 8.314            # [J/(mol·K)]
delta_Hr = -50000    # [J/mol] (exothermic)
Q_max = 2            # [kW]
CA_feed = 2.0        # [mol/L]
T_min = 300          # [K]
T_max = 400          # [K]


def production_rate(x):
    """Objective: Maximize r_B = k(T) * C_A^2."""
    C_A, T = x
    k = k0 * np.exp(-Ea / (R * T))
    return -k * C_A**2  # Negative for maximization


def cooling_constraint(x):
    """Constraint: r_B <= Q_max / (-ΔH_r) (in mol/L·s)."""
    C_A, T = x
    k = k0 * np.exp(-Ea / (R * T))
```

```
    r_B = k * C_A**2
    max_r_B = (Q_max * 1000) / (-delta_Hr)  # Convert Q_max to J/s
    return max_r_B - r_B  # Must be >= 0

# Constraints
cons = [
    {'type': 'ineq', 'fun': lambda x: CA_feed - x[0]},  # C_A <=
CA_feed
    {'type': 'ineq', 'fun': lambda x: x[1] - T_min},    # T >= T_min
    {'type': 'ineq', 'fun': lambda x: T_max - x[1]},    # T <= T_max
    {'type': 'ineq', 'fun': cooling_constraint}         # r_B <=
Q_max / (-ΔH_r)
]

# Initial guess [C_A, T]
x0 = np.array([1.0, 350])

# Solve
result = minimize(production_rate, x0, method='SLSQP',
constraints=cons)

# Results
C_A_opt, T_opt = result.x
k_opt = k0 * np.exp(-Ea / (R * T_opt))
r_B_opt = k_opt * C_A_opt**2
Q_generated = (-delta_Hr * r_B_opt) / 1,000 # [kW]

print(f"Optimal C_A: {C_A_opt:.2f} mol/L")
print(f"Optimal T: {T_opt:.2f} K")
print(f"Max production rate: {r_B_opt:.4f} mol/(L·s)")
print(f"Heat generated: {Q_generated:.2f} kW (<= Q_max = {Q_max} kW)")
```

We will find:

```
Optimal C_A: 1.93 mol/L
Optimal T: 300.00 K
Max production rate: 2.50 mol/(L·s)
Heat generated: 2.00 kW (<= Q_max = 2 kW)
```

To visualize the constraints and the optimum in a contour plot, by adding the following code to our script:

```
import matplotlib.pyplot as plt

# Generate grid
C_A_vals = np.linspace(0.5, 2.0, 100)
T_vals = np.linspace(300, 400, 100)
C_A_grid, T_grid = np.meshgrid(C_A_vals, T_vals)
r_B = np.zeros_like(C_A_grid)
Q = np.zeros_like(C_A_grid)

for i in range(100):
    for j in range(100):
        r_B[i,j] = k0 * np.exp(-Ea / (R * T_grid[i,j])) * C_A_grid
[i,j]**2
        Q[i,j] = -delta_Hr * r_B[i,j] / 1000

plt.contourf(C_A_grid, T_grid, r_B, levels=20, cmap='viridis')
plt.colorbar(label='Production rate (mol/L·s)')
plt.contour(C_A_grid, T_grid, Q, levels=[500], colors='red',
linewidths=2)
plt.plot(C_A_opt, T_opt, 'ro', label='Optimal point')
plt.xlabel('C_A (mol/L)'); plt.ylabel('T (K)')
plt.legend()
plt.show()
```

This leads to Figure 5.1.

The cooling constraint drives the solution, and it depends on the reaction rate r_B, which is also the objective. To maximize the objective, c_A and T should be as large as possible (toward their bounds), but the cooling constraint limits the temperature.

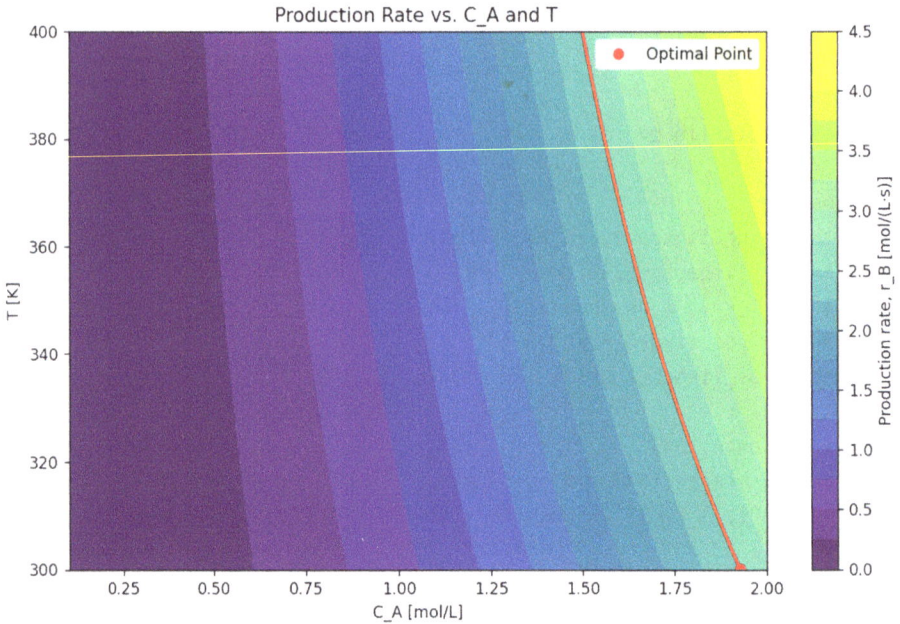

Figure 5.1: The decision space for T and c_A, showing the constraint for the cooling and the optimum point.

5.3 Solving NLP Problems

5.3.1 Unconstrained NLP

To find the optimum of an unconstrained NLP, we have to evaluate where the (partial) derivatives of the objective function, with respect to the decision variables, are zero. If it is a small NLP, we might calculate the derivatives explicitly leading to a set of algebraic equations which can be solved with Newton's method (see Chapter 2). If the computation of the derivatives is expensive, the derivatives might be estimated via finite differences and an adaptation of Newton's method (Broyden's method can be applied).

5.3.2 Constrained NLP

Direct Substitution: One way of solving NLP's is via *direct substitution*. This is only suitable for small problems. In many problems, elimination of a single equality constraint is superior to an approach in which the constraint is retained and some constrained optimization procedure is executed. In figure 5.2. below we can see John, the Phd student using NLP in solving food product formulation problems.

Suppose we have the following problem:

Figure 5.2: John, the PhD student, is using NLP to optimize a food product formulation.

$$\min f(x_1, x_2) = 4x_1^2 + 5x_2^2$$

subject to:

$$2x_1 + 3x_2 = 6$$

either x_1 or x_2 could be eliminated. Solving for x_1, we can rewrite the constraint:

$$x_1 = \frac{6 - 3x_2}{2}$$

And then plug it into the objective function:

$$f(x_2) = 14x_2^2 - 36x_2 + 36$$

The objective function is now only dependent on x_2 and the extremum can be found by solving the following equation:

$$\frac{df}{dx_2} = 28x_2 - 36 = 0 \quad x_2^* = 1.286$$

The value for x_1 can now be found by filling in x_2^* into the constraint:

$$x_1^* = \frac{6 - 3x_2^*}{2} = 1.071$$

In problems with n variables and m equality constraints, we can eliminate m variables by direct substitution. The objective function then has to be differentiated with respect to the remaining $(n-m)$ variables, where all partial derivatives are set to zero.

If the objective function is convex and the constraints are for a convex region (as in the example), then the stationary point is a global minimum. However, often problems are not coming this simple.

Lagrange Multipliers: In the *Lagrange multiplier method*, we combine the objective and constraints into a *Lagrangian function*. This Lagrangian function can be obtained by looking at the first-order necessary conditions for a local extremum.

Suppose we want to solve the following NLP:

$$\min f(x_1, x_2) = x_1 + x_2$$

subject to:

$$h(x_1, x_2) = x_1^2 + x_2^2 - 1 = 0$$

This problem is illustrated in Figure 5.3.

The contours of the linear objective are lines parallel to the one in the figure. The contour of the lowest value touches the circle at the optimum point $x^* = (-0.707, -0.707)$. If we look at the gradient of the objective function and the constraint at the optimum point, we find the following equation:

$$\nabla f(x^*) = [1, 1]$$

and

$$\nabla h(x^*) = [2x_1, 2x_2]_{x*} = [-1.414, -1.414]$$

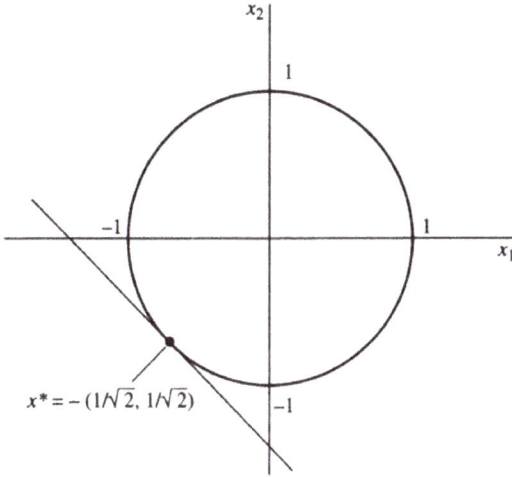

Figure 5.3: Circular feasible region with objective function contours and constraints.

The gradient of the objective function is *orthogonal* to the tangent plane of the constraint at x^*. Generally, the gradient of the constraint is always orthogonal to this tangent plane at the optimum point. In other words, the gradient of the objective function and the constraint are *collinear*, that is, they lie on the same line, but point in opposite directions. This means the two vectors must be multiples of each other:

$$\nabla f(x^*) = \lambda^* \nabla h\ (x^*) \tag{5.1}$$

where for our example $\lambda^* = -0.707$. λ is called the *Lagrange multiplier* for the constraint $h = 0$. The equation above can be rewritten as follows:

$$\nabla f(x^*) + \lambda^* \nabla h(x^*) = 0 \tag{5.2}$$

with $\lambda^* = 0.707$. We now introduce a new function $\mathcal{L}(x, \lambda)$ which is called the *Lagrangian function*:

$$\mathcal{L}(x, \lambda) = f(x) + \lambda h(x) \tag{5.3}$$

Now eq. (2) becomes:

$$\nabla(x, \lambda)|_{x*, \lambda*} = 0 \tag{5.4}$$

The gradient of the Lagrangian function with respect to x, evaluated at the optimum x and λ, is zero, plus the feasibility condition:

$$h(x^*) = 0 \tag{5.5}$$

constitutes the first-order necessary conditions for optimality.

For our example, we can now write the Lagrangian function as follows:

$$\mathcal{L}(x,\lambda) = (x_1 + x_2) + \lambda\left(x_1^2 + x_2^2 - 1\right)$$

Setting the partial derivatives of the Lagrangian function to zero leads to the following:

$$\frac{\partial \mathcal{L}}{\partial x_1} = 1 + 2\lambda x_1 = 0$$

$$\frac{\partial \mathcal{L}}{\partial x_2} = 1 + 2\lambda x_2 = 0$$

And the feasibility condition:

$$\frac{\partial \mathcal{L}}{\partial \lambda} = \left(x_1^2 + x_2^2 - 1 = 0\right)$$

We have now obtained three equations with three unknowns, which can be solved to find $x_1^* = x_2^* = \pm 0.707$ and $\lambda^* = \pm 0.707$

A general equality-constrained NLP with m constraints and n variables can be written as follows:

$$\max f(x)$$

subject to:

$$h_j(x) = b_j, \quad j = 1, \ldots, m$$

where $x = (x_1, \ldots, x_n)$ are the decision variables and b_j are the right-hand sides of the equality constraints. For this problem, the Lagrangian function can be written as follows:

$$\mathcal{L}(x,\lambda) = f(x) + \sum_{j}^{m} \lambda_j \left[h(x) - b_j\right]$$

And the first-order necessary conditions are as follows:

$$\frac{\partial \mathcal{L}}{\partial x_i} = \frac{\partial f}{\partial x_i} + \sum_{j}^{m} \lambda_j \frac{\partial h_j}{\partial x_i} = 0, \quad i = 1, \ldots, n$$

$$h_j(x) = b_j, \quad j = 1, \ldots, m$$

It is noted that local optima also satisfy the first-order necessary conditions and that even a saddle point or inflection point can be found as a solution.

In a similar way, NLP with only inequality constraints can be solved. But the reality is that NLP often have equality constraints as well as inequality constraints.

Let us now consider a general NLP with m equality constraints, r inequality constraints, and n variables:

$$\max f(x)$$

subject to:

$$h_i(x) = b_i \quad j = 1, \ldots, m$$

$$g_j(x) \le c_j, \quad j = 1, \ldots, r$$

We will now write the Lagrangian function as follows:

$$\mathcal{L}(x, \lambda, v) = f(x) + \sum_i^m \lambda_i [h(x) - b_i] + \sum_j^r v_j [g_j(x) - c_j]$$

Now x^* is a local minimum if there exist x and multipliers (λ and v) such that:

$$\nabla \mathcal{L}(x^*, \lambda^*, v^*) = \sum_i^m \lambda_i^* \nabla h_i(x^*) + \sum_j^r v_j^* \nabla g_j(x^*) = 0$$

In this case, we need *complementary slackness* for the inequalities:

$$u_j^* \ge 0 \; u_j^* [g_j(x^*) - c_j] = 0, \quad j = 1, \ldots, r$$

These conditions are also called the *Karush-Kuhn-Tucker (KKT) conditions for optimality*. Let us try this with an example.

Example: The goal is to maximize the production rate P of a chemical reaction in a continuous stirred-tank reactor (CSTR), where P is a function of x and y (concentrations of reactants):

$$\max P(x, y) = 2x^2 + 3y^2 - xy$$

There are two constraints: an equality constraint for the total mass balance (feed stream allocation):

$$x + y = 10$$

and an inequality that defines safety limits (to avoid explosive conditions):

$$x^2 + y^2 \le 52$$

We can formulate the Lagrangian function:

$$\mathcal{L}(x, y, \lambda, v) = 2x^2 + 3y^2 - xy + \lambda(10 - x - y) + v(50 - x^2 - y^2)$$

From the necessary conditions of optimality follows:

$$\frac{\partial \mathcal{L}}{\partial x} = 4x - y - \lambda - 2vx = 0$$

$$\frac{\partial \mathcal{L}}{\partial y} = 6y - x - \lambda - 2vy = 0$$

and the complementary slackness:

$$v \geq 0 \quad v\left(52 - x^2 - y^2\right) = 0$$

We can now solve this problem for two cases. In case the inequality constraint is inactive ($v = 0$):

From the derivatives follows:

$$4x - y - \lambda = 0 \text{ and } 6y - x - \lambda = 0$$

We subtract to eliminate λ:

$$5x - 7y = 0 \quad x = \frac{7}{5}y$$

Substitute in the equality constraint:

$$\frac{7}{5}y + y = 10$$

We find that $y^* = \frac{25}{6} \approx 4.17$ and $x^* = \frac{35}{6} \approx 5.83$.

The Lagrange multiplier l can be found by the substitution of x^* and y^* into the derivative, and we will find $\lambda = \frac{115}{6} \approx 19.17$ (which should be positive).

We can do a constraint check:

$$x^2 + y^2 = \left(\frac{35}{6}\right)^2 + \left(\frac{25}{6}\right)^2 = \frac{1,850}{36} \approx 51.39 \leq 52$$

which is feasible, but not necessarily optimal. What would we find if the inequality is active? That means that v is not zero, it has to be positive.

From the derivatives now follows:

$$4x - y - \lambda - 2vx = 0$$

$$6y - x - \lambda - 2vy = 0$$

We can rewrite the equality constraint $y = 10 - x$ and substitute it in the inequality constraint:

$$x^2 + (10 - x)^2 = 52 \rightarrow 2x^2 - 20x + 48 = 0$$

This leads to a value for $x = 4$ or $x = 6$.

Let us start with the first candidate solution with $x = 4$, we will find from the equality constraint that $y = 10 - x = 4$

We can now find via the derivatives the Lagrange multipliers:

$$16 - 6 - \lambda - 8v = 0 \rightarrow 10 - \lambda - 8v = 0$$

$$36 - 4 - \lambda - 12v = 0 \rightarrow 32 - \lambda - 12v = 0$$

These are two equations with two unknowns, which solve to: $v = 5.5$ and $\lambda = -34$.

For the second candidate solution, where $x = 6$ and $y = 4$, we will find the multiplier $v = -0.5$, which is invalid, because it has to be a positive number.

We can verify the solution numerically with Python:

```
from scipy.optimize import minimize
def objective(z):
    x, y = z
    return -(2*x**2 + 3*y**2 - x*y)  # Negative for maximization

cons = [
    {'type': 'eq', 'fun': lambda z: z[0] + z[1] - 10},
    {'type': 'eq', 'fun': lambda z: z[0]**2 + z[1]**2 - 52}  #
Enforce active inequality
]
result = minimize(objective, x0= [5, 5], constraints=cons,
method='SLSQP')
x_opt, y_opt = result.x
print(f"Optimal x: {x_opt:.0f}, y: {y_opt:.0f}")
print(f"Production rate: {-result.fun:.0f}")
```

which gives as an output:
```
Optimal x: 4, y: 6
Production rate: 116
```

We can also visualize the result by adding the following code to the script:

```
import matplotlib.pyplot as plt

# Feasible region
theta = np.linspace(0, 2*np.pi, 100)
x_circle = np.sqrt(52) * np.cos(theta)
y_circle = np.sqrt(52) * np.sin(theta)

# Constraint line
x_line = np.linspace(0, 10, 100)
y_line = 10 - x_line
# Plot
```

```
plt.figure(figsize=(8, 6))
plt.plot(x_circle, y_circle, 'r-', label=r'$x^2 + y^2 \leq 52$')
plt.plot(x_line, y_line, 'b-', label=r'$x + y = 10$')
plt.plot(x_opt, y_opt, 'ko', label='Optimal Point')
plt.xlabel('x'); plt.ylabel('y')
plt.legend(); plt.grid()
plt.title('Feasible Region and Optimal Solution')
plt.axis('equal')
plt.show()
```

This leads to Figure 5.4.

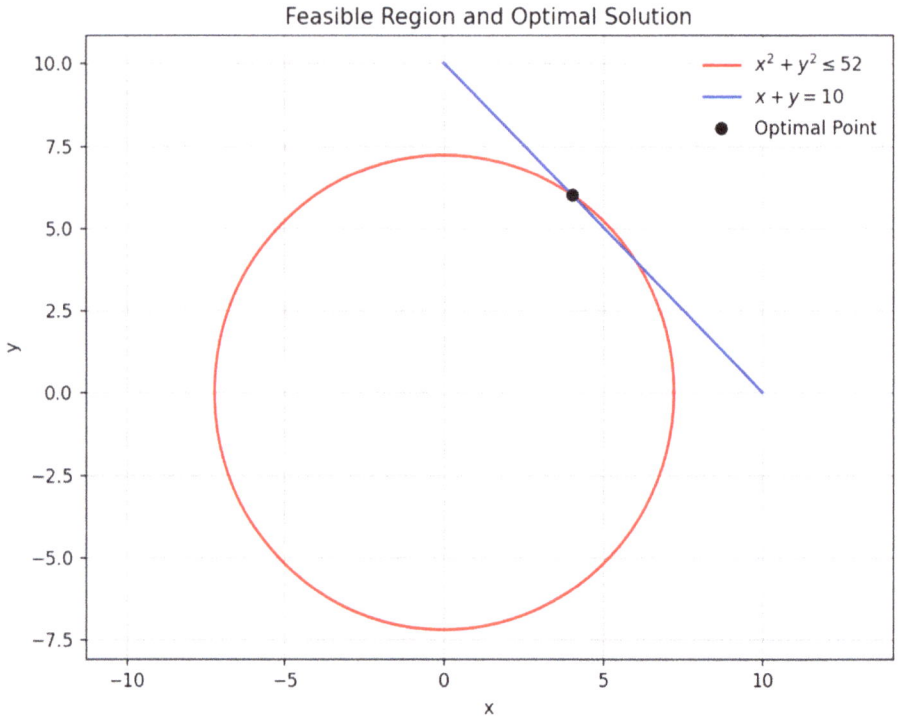

Figure 5.4: Constraints and optimal point.

It turns out that the Lagrange multipliers λ and ν correspond with the shadow prices for the equality constraint and the inequality constraint, respectively.

Successive Linear Programming (SLP): In SLP, we linearize the original problem on the basis of a Taylor series expansion and subsequently solve linear subproblems in a successive way.

From the Expert: **Prof. Johan Grievink**
Delft University of Technology, The Netherlands

Balancing Between Context and Contents in (Industrial) Optimization Applications
An (industrial) chemical engineering optimization effort refers to an object that (will) exist in a physical reality and has one or more performance characteristics. One can optimize the performance(s) of such an object by mathematical and computational means. There will be distinctive relationships between an object and its computational model. A model functions in the *context* of its object. The model itself has *contents* in terms of mathematical equations and variables, its digital implementation, and a numerical solver. Both context and contents play roles in a workflow for optimization projects (Figure 1).

A workflow for chemical engineering optimization applications

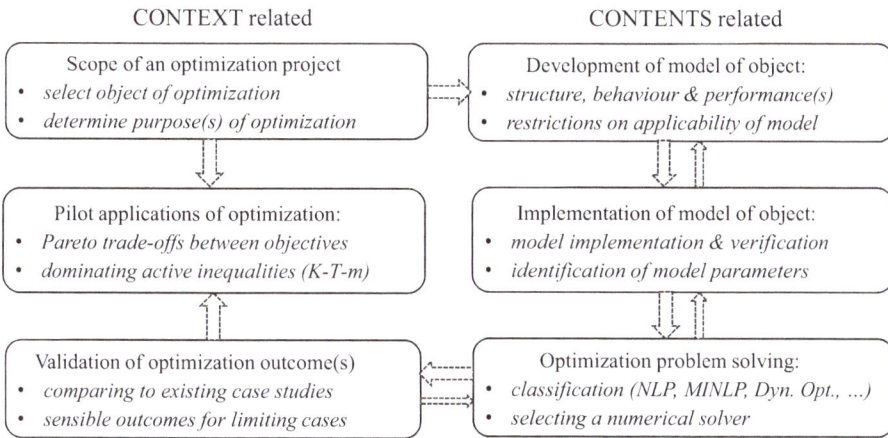

CONTEXT related   CONTENTS related

Scope of an optimization project
- *select object of optimization*
- *determine purpose(s) of optimization*

Development of model of object:
- *structure, behaviour & performance(s)*
- *restrictions on applicability of model*

Pilot applications of optimization:
- *Pareto trade-offs between objectives*
- *dominating active inequalities (K-T-m)*

Implementation of model of object:
- *model implementation & verification*
- *identification of model parameters*

Validation of optimization outcome(s)
- *comparing to existing case studies*
- *sensible outcomes for limiting cases*

Optimization problem solving:
- *classification (NLP, MINLP, Dyn. Opt., ...)*
- *selecting a numerical solver*

Figure 1: Relationship between context and contents in chemical engineering optimization.

The left-hand side of Figure 1 deals with the *context*, i.e., the relationships between the (real life) object of optimization and its computational model. That is, how well do a problem statement, a model, and the obtained optimization solutions mimic a real-life optimization situation of the chemical engineering object. The right-hand side in the figure deals with the mathematical and computational *contents* of the optimization model. The capabilities to create extensive contents are growing tremendously because of the continuing advancements in digital technologies. This presents a challenge for keeping a proper functional balance between context and contents.

For optimizing the performance of a chemical engineering object, one needs contextual information for the object model. It creates a frame for model content needed for the delivery of matching results. Reversely, model-based explorations of the performance space, based on strong model content, may reveal new options for better design and operations of the object of interest. Another contextual item is the observance of the domain of validity of a chemical engineering model. Optimizers search over the full space of variables and may arrive at conditions where the validity of some empirical model equations is questionable. Delineating the domains of validity of empirical correlations by inequality constraints avoids entering terra incognito. In a similar contextual vein, it is useful to understand the nature of the optimum: Which inequality constraints are active at the optimum? One can list the computed Kuhn-Tucker multipliers of the inequality constraints in decreasing order of magnitude. Knowing this, one may consider options for shifting the severest inequality constraints to more lenient values. It widens the feasible domain while increasing performance. Such a potential increase in performance may even offset the costs of shifting the severest inequality constraints to more lenient values. In view of the growing overpotential for computational contents, the contextual aspects will become more critical for the practical success of optimization studies.

Let us demonstrate SLP with a numerical example. Suppose I have the following NLP, with linear objective function and two nonlinear inequalities:

$$\min 2x + y$$

subject to:

$$x^2 + y^2 \leq 25$$

$$x^2 - y^2 \leq 7$$

$$x, y \geq 0$$

We will start with a guess value, for example, $(x_C, y_C) = (2, 2)$. Around this point, we will linearize the NLP with a *Taylor series*:

$$\min 2x_C + 2\Delta x + y_C + \Delta y$$

subject to:

$$x_C^2 + 2x_C\Delta x + 2y_C^2 + 2y_{C\Delta y} \leq 25$$

$$x_C^2 + 2x_C\Delta x - y_C^2 - 2y_C\Delta y \leq 7$$

$$2 + \Delta x \geq 0, 2 + \Delta y \geq 0$$

We could pose bounds on the steps, to ensure the errors between the nonlinear problem and the linearized model are not too large:

$$-1 \leq \Delta x \leq 1, \; -1 \leq \Delta y \leq 1$$

After filling in the values for $(x_C, y_C) = (2, 2)$, the problem further simplifies to:

$$\min 2\Delta x + \Delta y$$

subject to:

$$\Delta x + \Delta y \le 4.25$$

$$\Delta x - \Delta y \le 1.75$$

$$-1 \le \Delta x \le 1, \ -1 \le \Delta y \le 1$$

We can solve this LP and find: $(\Delta x, \Delta y) = (1, 1)$. With these deltas, we calculate new guess values $(x_{C,new}, y_{C,new}) = (x_C + \Delta x, y_C + \Delta y) = (3, 3)$. And around this new point, we linearize again and solve the resulting LP. For the second iteration, we find $(\Delta x, \Delta y) = (1, \frac{1}{6})$. With these deltas, we define a set of new guess values $(x_{C,new}, y_{C,new}) = (x_C + \Delta x, y_C + \Delta y) = (4, 3.167)$, which is already close to the actual optimum (4,3). With one more iteration, we find (4,3.005) as solution. The successive movement of the optimum during the iterations is illustrated in Figure 5.5.

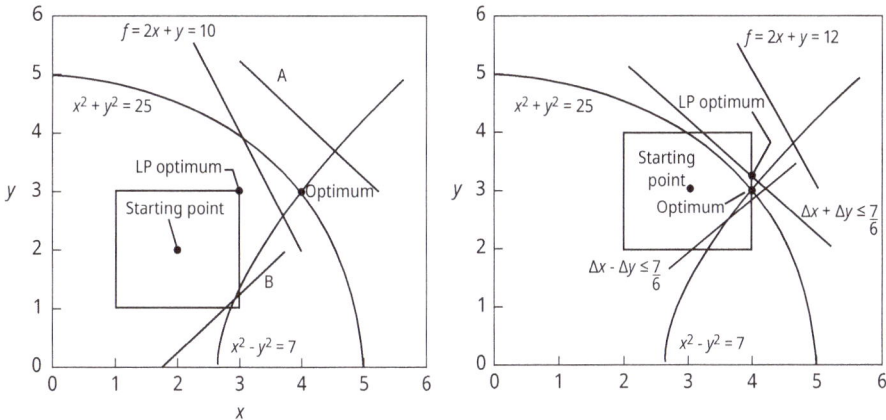

Figure 5.5: SLP. Left: From initial guess (2,2) to (3,3) in the first iteration. Right from (3,3) to (4,3.167) in the second iteration. Already close to the actual optimum (4,3).

Quadratic Programming: *Quadratic programming* (QP) is a specialized form of mathematical optimization that plays a pivotal role in chemical engineering design and operations. At its core, QP seeks to minimize or maximize a quadratic objective function subject to linear constraints, making it more flexible than LP while remaining computationally tractable. The general formulation involves an objective function that is quadratic in the decision variables, typically expressed as follows:

$$\min \frac{1}{2} x^T Q x + c^T x$$

where x represents the vector of decision variables, Q is a symmetric matrix defining the quadratic terms, and c captures the linear components. This structure allows QP to model phenomena with inherently quadratic relationships, such as kinetic energy in mechanical systems or variance in portfolio optimization.

A defining feature of quadratic programs is their constraint set, which is restricted to linear equalities and inequalities. These constraints often represent mass balances, energy limits, or safety thresholds in chemical processes. For instance, in reactor design, a QP might optimize catalyst distribution while respecting feedstock availability constraints expressed as follows:

$$Ax \leq b$$

The quadratic objective could then represent the trade-off between reaction yield and energy consumption. The elegance of QP lies in its ability to model such trade-offs precisely while leveraging efficient algorithms.

The solvability of a QP depends critically on the properties of the Q matrix. When Q is positive semidefinite, the problem is convex, guaranteeing that any local minimum is also global – a property that simplifies solution strategies. Convex QPs arise naturally in areas such as least-squares regression and process control, where objectives often involve minimizing squared errors or deviations. In contrast, non-convex QPs, where Q has negative eigenvalues, introduce computational challenges, including the possibility of multiple local optima. Such problems require advanced techniques such as branch-and-bound or spatial branching, though they are less common in standard engineering applications.

From an algorithmic perspective, quadratic programs benefit from well-established solution methods. Active-set methods excel for small- to medium-scale problems by iteratively identifying which constraints are "active" at the optimum. Interior-point methods, on the other hand, are preferred for large-scale QPs due to their polynomial-time complexity and ability to handle dense constraint matrices. Modern solvers such as CPLEX and Gurobi integrate these approaches, often achieving solutions rapidly even for problems with thousands of variables.

In chemical engineering practice, QP finds widespread use in model predictive control, where it optimizes process trajectories while respecting equipment limits. For example, a distillation column's operating conditions might be adjusted in real time by solving a QP that minimizes energy use (a quadratic function of flow rates) subject to purity specifications (linear constraints). The quadratic objective naturally penalizes large deviations from set points, leading to smoother control actions compared to linear objectives.

The connection between QP and fundamental engineering principles extends to sensitivity analysis. The Lagrange multipliers associated with constraints quantify how the optimal objective value changes with respect to constraint relaxations – a critical insight for debottlenecking processes. Moreover, the dual problem of a QP provides a lower bound on the objective, offering a way to assess solution quality and diagnose infeasibilities.

Despite its utility, QP has limitations. The requirement for linear constraints can be restrictive when modeling nonlinear phenomena, necessitating approximations or extensions like sequential quadratic programming (SQP). Additionally, memory demands grow quadratically with problem size due to the Q matrix, though sparse matrix techniques mitigate this for structured problems.

In summary, quadratic programming occupies a vital niche in process optimization, combining the precision of quadratic modeling with the robustness of linear constraints. Its theoretical guarantees, efficient algorithms, and adaptability to engineering needs – from plant design to real-time control – make it an indispensable tool for chemical engineers tackling complex, constrained optimization challenges. Mastery of QP empowers practitioners to navigate trade-offs systematically, ensuring optimal performance while adhering to physical and operational limits.

Sequential Quadratic Programming (SQP): It stands as one of the most powerful and widely used methods for solving *nonlinear optimization problems* with constraints, particularly in chemical engineering applications where processes exhibit inherently nonlinear behavior. At its core, SQP solves complex nonlinear problems by iteratively approximating them as quadratic programming subproblems – a strategy that marries the efficiency of QP solvers with the flexibility needed to handle nonlinear objectives and constraints. The method's popularity stems from its robust convergence properties and ability to handle large-scale systems, from reactor networks to refinery scheduling.

The essence of SQP lies in its iterative approach. For each iteration, the algorithm constructs a local quadratic approximation of the Lagrangian function, which incorporates both the original nonlinear objective and constraints via Lagrange multipliers. This approximation transforms the nonlinear problem into a QP subproblem that is easier to solve. The subproblem's solution then provides a search direction, which is combined with a step-size strategy (like line search or trust-region methods) to update the current design variables. By repeating this process, SQP refines the solution until it converges to a point satisfying the KKT conditions – the gold standard for optimality in constrained optimization.

A key strength of SQP is its ability to handle *nonlinear equality and inequality constraints*, which are ubiquitous in engineering. For example, in optimizing a distillation column, SQP can navigate nonlinear vapor-liquid equilibrium (VLE) relationships (equality constraints) while respecting bounds on temperature and pressure (inequality constraints). The method linearizes these constraints at each iteration but retains their nonlinearity in the overall convergence path, ensuring feasibility without oversimplifying the physics. This balance between accuracy and computational tractability makes SQP indispensable for problems where linearizations alone would fail.

The algorithm's efficiency hinges on two components:
1. **QP Subproblem Construction**: The Hessian of the Lagrangian (second-order derivatives) is approximated – often using quasi-Newton methods such as Broyden-

Fletcher-Goldfarb-Shanno (BFGS) – to maintain sparsity and reduce computational cost.
2. **Globalization Strategy**: Trust regions or line searches ensure progress toward the optimum even when far from the solution, preventing divergence.

In practice, SQP excels in scenarios such as *real-time optimization* of chemical plants, where models must adapt to changing feedstocks or market demands. Consider a plug-flow reactor where the goal is to maximize yield while avoiding hot spots. SQP can dynamically adjust operating conditions by solving a sequence of QP problems, each accounting for updated measurements and constraints. This adaptability is critical for processes where safety and economics are tightly coupled.

However, SQP is not without challenges. The need for accurate second-derivative information can be computationally expensive, though modern implementations often use limited-memory BFGS to alleviate this. Additionally, the method assumes smoothness in objective and constraint functions, which may not hold for problems with discontinuities (e.g., phase changes). For such cases, hybrid approaches combining SQP with heuristic methods are employed.

The convergence behavior of SQP is typically *superlinear* near the solution, meaning it outperforms gradient-based methods in later iterations. This property is particularly valuable in *multidisciplinary design optimization*, where variables from thermodynamics, kinetics, and control interact nonlinearly. By leveraging curvature information, SQP efficiently navigates these interactions, avoiding the pitfalls of slower first-order methods.

In summary, SQP bridges the gap between theoretical optimization and industrial practicality. Its iterative QP framework, coupled with robust globalization strategies, makes it a cornerstone for solving nonlinear problems in chemical engineering – from catalyst design to plant-wide optimization. While it demands careful implementation, its ability to deliver fast, reliable solutions to highly constrained problems ensures its enduring relevance in the field.

Penalty Methods: The main idea of a *penalty method* is to convert a nonlinear program with constraints to a sequence of unconstrained problems.

As an example, consider the following:

$$\min f(x_1, x_2) = (x_1 - 1)^2 + (x_2 - 2)^2$$

subject to:

$$h(x_1, x_2) = x_1 + x_2 - 4 = 0$$

We formulate a new unconstrained objective function:

$$\min P(x_1, x_2, r) = (x_1 - 1)^2 + (x_2 - 2)^2 + r(x_1 + x_2 - 4)$$

The constraint has been added to the objective function with a *penalty factor* (or *weight factor*) r. We cannot solve this unconstrained optimization model for different values of r. The outcomes can be found in Table 5.2.

Table 5.2: Effect of penalty weighting r on minimum of f.

r	x_1	x_2	f
0	1.0000	2.0000	0.0000
0.1	1.0833	2.0833	0.0833
1	1.3333	2.3333	0.3333
10	1.4762	2.4762	0.4762
100	1.4975	2.4975	0.4975
1,000	1.4998	2.4998	0.4998
x^*	**1.5000**	**2.5000**	**0.5000**

It is clear that by increasing the value of r, the solutions of x_1, x_2, and f converge. Figure 5.6 visualizes what happens with the objective function for different values of r.

But for large r, the Hessian of P is called, *ill-conditioned*, and it becomes more difficult for a solver to minimize P.

The *condition number* can be used to estimate the condition of P. The condition number is defined as the ratio of the largest and smallest eigenvalues of the Hessian of P. If $P = 1$ = small, $P = 10^5$ = moderately large, $P = 10^9$ = large, and $P = 10^{14}$ = extremely large.

By moving the constraints into the objective function, we *relax* the problem. That is why we sometimes speak of a *Lagrange relaxation*, as an analog to the definition of the Lagrangian function.

Barrier Methods: Like penalty methods, *barrier methods* convert a constrained optimization problem to a series of unconstrained ones. The solutions to these subproblems are at the interior of the feasible region and converge to the constrained solution as positive barrier parameters go to zero. This approach contrasts with the penalty method, whose unconstrained subproblems converge from outside the feasible region.

Considering the same example as presented with the penalty method, we could write a *logarithmic barrier function*:

$$\min B(x_1, x_2, r) = (x_1 - 1)^2 + (x_2 - 2)^2 - r \ln(x_1 + x_2 - 4)$$

While r is decreasing, the term $-r \ln(g(x_1, x_2))$ goes to infinity, creating an infinitely high barrier along this boundary.

Smoothed Method: The *smoothed optimization method* is useful when the problem has soft constraints, e.g., certain costumer demands or capacity limits that are allowed to be (modestly) violated.

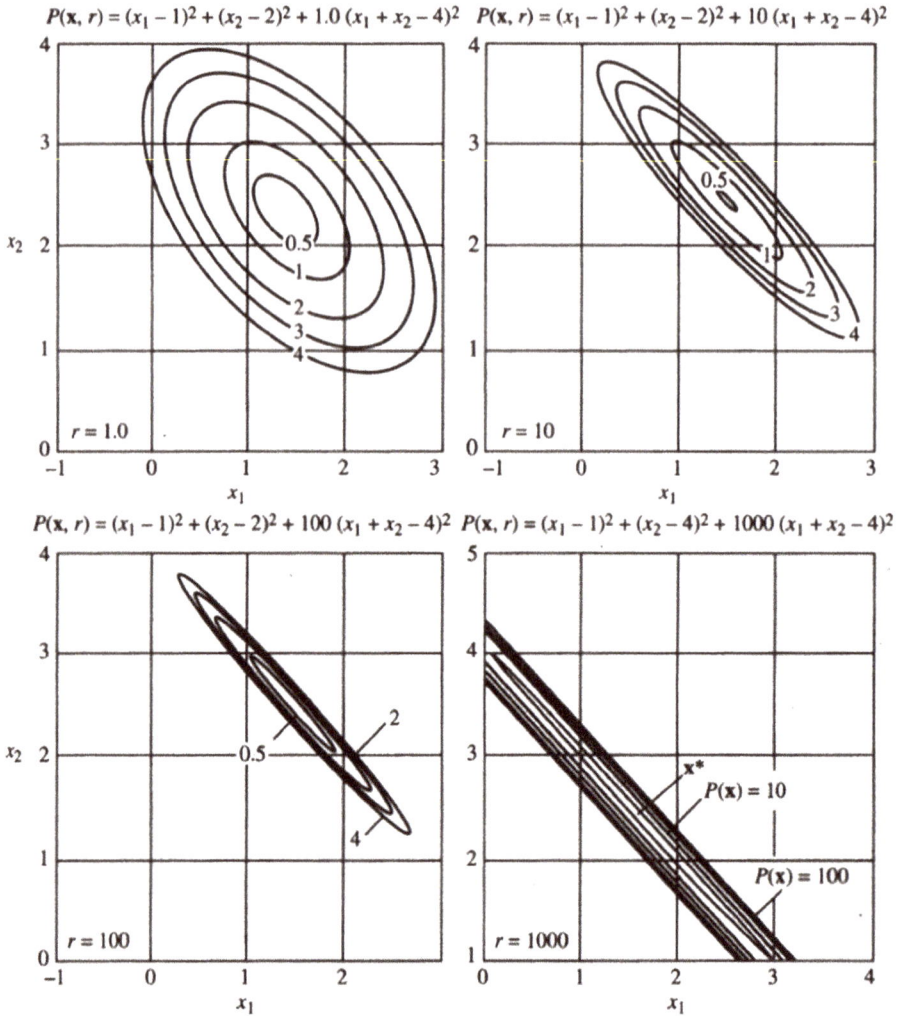

$P(\mathbf{x}, r) = (x_1 - 1)^2 + (x_2 - 2)^2 + 1.0\,(x_1 + x_2 - 4)^2$

$P(\mathbf{x}, r) = (x_1 - 1)^2 + (x_2 - 2)^2 + 10\,(x_1 + x_2 - 4)^2$

$P(\mathbf{x}, r) = (x_1 - 1)^2 + (x_2 - 2)^2 + 100\,(x_1 + x_2 - 4)^2$

$P(\mathbf{x}, r) = (x_1 - 1)^2 + (x_2 - 4)^2 + 1000\,(x_1 + x_2 - 4)^2$

Figure 5.6: Transformation of a constrained problem to an unconstrained equivalent problem. The contours of the unconstrained penalty function are shown for different values of r.

Suppose we have the following NLP:

$$\min f(x)$$

subject to:

$$h_i(x) = 0 \quad i = 1, \ldots, m$$

$$g_j(x) = 0 \quad j = 1, \ldots, r$$

We can rewrite this NLP to its *elastic formulation*:

$$\min f(x) + \sum_{i}^{m} w1_i(p1_i + n1_i) + \sum_{j}^{r} w2_j p2_j.$$

Subject to:

$$h_i(x) = p1_i - n1_i \ i = 1, \ldots, m$$

$$g_j(x) = p2_j - n2_j \ j = 1, \ldots, r$$

$$p1_i, \ p2_j n1_i, \ n2_j \geq 0$$

The p's are positive deviation variables, and the n's are negative deviation variables. If the p's and n's take values, it effectively means that the constraints are violated. By modifying the weights w's in the objective function, we can stir how much violation is accepted.

5.4 Challenges in NLP

In solving NLP's, we can encounter several challenges. Continuity and convexity of objective function and constraints are critical. If we are dealing with a non-convex problem, we might end up with finding a local optimum, instead of a global one. Sometimes, multi-start algorithms or metaheuristics are employed to systematically scan the decision space. Non-convex problems might be convexified, that is, reformulated in such way that the problem becomes convex. To learn more about convexity, continuity, and local/global extrema, the reader is referred to Chapter 2.

5.5 Exercises

Exercise 1: Reactor Temperature Optimization ★
Problem:
A CSTR has a reaction rate $r = k_0 e^{\frac{E_a}{RT}} c_A$ where $k_0 = 10^7$ s^{-1}, $E_a = 75$ kJ/mol, and the feed concentration CA = 2.0 mol/L.
- **Objective**: Maximize r by optimizing T (300 K $\leq T \leq$ 500 K).
- **Constraint**: The cooling system limits $T \leq 450$ K.

Tasks:
1. Formulate the NLP.
2. Solve graphically or using Python.

Hint: The objective is unimodal; check derivatives at bounds.

Exercise 2: Nonlinear Constraint Handling ★★

Problem:

Minimize energy cost: $f(T) = 5T^3 - 100T^2 + 500T$ for a distillation column, subject to:
- **Equality Constraint**: VLE $T - 0.2y^2 = 350$ (y = vapor fraction).
- **Inequality Constraint**: K300 ≤ T ≤ 400 K.

Tasks:
1. Rewrite the equality constraint to eliminate y.
2. Use Lagrange multipliers to find the optimum.

Hint: Substitute $y = \frac{T - 350}{0.2}$ into $f(T)$.

Exercise 3: Parameter Estimation (Nonlinear Regression) ★★

Problem:

Fit the nonlinear model $y = k_1 e^{k_2 t} t$ to data:

T (min)	1	2	3
Y (g/L)	1.8	1.2	0.8

Tasks:
1. Formulate the least-squares objective min $\sum$(ymodel – ydata)2 min$\sum$(ymodel – ydata)2.
2. Solve using `scipy.optimize.curve_fit`.

Hint: Logarithms can linearize the model for initial guesses.

Exercise 4: Multi-objective Optimization ★★★

Problem:

A reactor design has two conflicting objectives:
1. Maximize conversion $X = 1 - e^{-kT}$ (k = 0.1 s − 1k = 0.1 s − 1).
2. Minimize energy cost $C = 0.5T^2$

Tasks:
1. Plot the Pareto frontier for ∈ [300,500]K.
2. Identify the knee point where dX/dC is steepest.

Hint: Vary T and compute (X,C) pairs.

Exercise 5: SQP for Non-convex Problem ★★★★

Problem:

Minimize $f(x,y) = x^4 - 5x^2 + y^2$

subject to:
- $x + y \geq 1$,

- $x^2 + y^2 \le 4$.

Tasks:
1. Identify local minima using KKT conditions.
2. Solve using `scipy.optimize.minimize(method = 'SLSQP')`.

Hint: Check $(x,y) = (0,0)(x,y) = (0,0)$, $(1.5,0.5)(1.5,0.5)$, and $(-1.5,2.5)(-1.5,2.5)$.

Exercise 6: Setting Up and Solving the Economic Dispatch Problem ★★★

Problem:

This power generation scheduling problem is formulated as an optimization problem. The objective of the optimization problem is stated as the minimization of the total operating cost. The problem is constrained by unit operating limits and a generation-load-loss balance. Mathematically, the problem is stated as follows:

$$\min C = \sum_i a_i + b_i P_i + c_i P_i^2,$$

subject to:

$$\sum_i P_i - P_{\text{load}} - P_{\text{loss}} - P_{\text{export}} = 0$$

$$P_i^L \le P_i \le P_i^U$$

where P_i is the amount of power produced by unit i in MW. a_i, b_i, and c_i are cost coefficients (in $/h, $/MWh, and $/MWh^2, respectively). The objective is to minimize the power generation costs (C) while producing enough power to satisfy the demand (P_{load}) while accounting for power production losses (P_{loss}) and potential export of power (P_{export}).

Suppose we have two power generators with the following characteristics:

Unit	Capacity (MW)	a_i ($/h)	b_i ($/MWh)	c_i ($/MWh^2)
1	180	115.2	28.7	0.025
2	315	265.6	25.1	0.012

The demand load is 382 MW, and it is reasonable to assume that the losses are 1% of the load. In our case, we will not produce power for export purposes.

Task:

Determine the optimal production plan for the economic dispatch problem formulated above by a) analytical method, using the Lagrange multiplier method and b) implementation in an optimization software.

1 Project Overview and Learning Objectives

In this one-week team project, you will formulate and solve a realistic NLP problem central to chemical process economics: the optimal design of a batch plant that produces multiple products.

By the end of this project, your team should be able to:

–   Formulate a complex, real-world engineering design problem as a constrained NLP,
–   Implement the NLP model in a suitable software environment (e.g., Python with scipy.optimize, MATLAB with fmincon, GAMS, or Excel Solver),
–   Analyze the results, including the optimal design, economic performance, and active constraints, and
–   Collaborate effectively to divide the tasks of model formulation, coding, sensitivity analysis, and report writing.

2 Problem Statement

Your company is planning a new batch plant to manufacture three specialty chemicals: Product A, Product B, and Product C. Each product is produced in a dedicated batch reactor, but all products share the same mixing and packaging unit (MPU).

The goal is to determine the optimal size (volume) of each batch reactor and the optimal size (processing rate) of the MPU to maximize the annual profit of the plant, considering equipment costs and operational time.

Given Data:

–   Product Requirements:
    – F_i = Annual production requirement for product i (kg/year):
        – F_A = 100,000
        – F_B = 200,000
        – F_C = 150,000
–   Processing Times:
    – t_{ij} = Processing time for product i in unit j (hours). The time in a reactor depends on its size (see below).
    –   Reaction Time: For a reactor of size V_i (in m³), the time (hr) to produce a batch of product i is: $t\_ireactor = \alpha_i * (V_i)^{\beta_i}$
        –   Product A: $\alpha_A = 8$, $\beta_A = -0.3$
        –   Product B: $\alpha_B = 10$, $\beta_B = -0.25$
        –   Product C: $\alpha_C = 12$, $\beta_C = -0.4$
    –   MPU Time: The time to process a batch in the MPU is fixed: $t\_impu = 2$ h for all products.
–   Economic Data:
    – S_i = Selling price of product i ($/kg):
        – $S_A = 50$

  - $S_B = 40$
  - $S_C = 60$
  - C_i = Raw material cost of product i ($/kg):
    - $C_A = 20$
    - $C_B = 15$
    - $C_C = 25$
- Equipment Cost:
  - The cost of a reactor ($) is: Cost_reactor = 50,000 * (V_i)^0.6.
  - The cost of the MPU ($) is: Cost_mpu = 80,000 * (R_mpu)^0.7, where R_mpu is the processing rate (kg/h).
- Plant Operation:
  - H = Total operational hours available per year = 7,920 h (330 days * 24 h).

3 Mathematical Formulation (Your NLP Task)

Your team must define the decision variables, objective function, and constraints.
- Decision Variables:
  1. V_A, V_B, V_C [m^3]: Sizes of the reactors for products A, B, and C.
  2. R_mpu [kg/h]: Processing rate of the mixing/packaging unit.
  3. B_i [kg/batch]: Batch size for product i. *(Hint: You may need to define this to calculate the number of batches).*
- Objective Function: Maximize Annual Profit
  Maximize: Z = (Total Revenue – Total Raw Material Cost) – Total Annualized Equipment Cost
  - Total Revenue = Σ (F_i * S_i)
  - Total Raw Material Cost = Σ (F_i * C_i)
  - Total Equipment Cost = (50,000*(V_A^0.6 + V_B^0.6 + V_C^0.6) + 80,000*(R_mpu^0.7)) / 3
    (Note: Dividing by 3 is a simple way to annualize the capital cost over a 3-year payback period).
- Constraints:
  1. Annual Production Requirements: The number of batches must be sufficient to meet demand. (F_i / B_i) * (t_ireactor + t_impu) ≤ H for each product i.
  2. MPU Capacity: The MPU must be able to handle the batch size within its time limit. B_i ≤ R_mpu * t_impu for each product i. (This links B_i and R_mpu)
  3. Reactor Size and Batch Size: The batch size must fit in the reactor. B_i / ρ_i ≤ V_i for each product i. (Assume density ρ_i = 900 kg/m^3 for all products).
  4. Logical Constraints: All decision variables must be positive. V_A, V_B, V_C, R_mpu, B_i > 0.

4 Deliverables and Schedule (1 Week)
- Day 1: Team formation and problem understanding. Discuss and finalize the mathematical model. *Divide tasks: Who will work on which part of the formulation?*
- Days 2–3: Implementation. Code the NLP problem in your chosen software. Begin solving and debugging.
- Day 4: Solution and sensitivity analysis. Find the optimal solution. Answer: What happens if the production requirement for Product B increases by 20%?
- Day 5: Prepare the final submission.

Final Submission (one PDF per team):
1. Team Members and Contribution: list of members and a brief description of each person's primary contribution.
2. Problem Formulation: A clear, typed presentation of your final NLP formulation (objective and constraints).
3. Results: A table of your optimal decision variables and the maximum profit.
4. Discussion:
    - Which constraints are "active" (binding) at the optimum? What does this tell you about the plant's limitations?
    - Briefly discuss the challenges you faced in solving the problem (e.g., initial guesses, scaling).
    - What was the result of your sensitivity analysis?
5. Appendix: A printout of your well-commented code.

5 Tips for Success
- Start Simple: Before coding the full problem, test your code with just one product to ensure the model works.
- Provide Bounds: Help the solver by providing reasonable lower and upper bounds for your variables (e.g., $10 < V_i < 200$, $100 < R\_mpu < 5{,}000$).
- Initial Guess: Choose sensible initial guesses. A bad guess can cause the solver to fail.
- Work Together: The model formulation is the hardest part. Do it together on a whiteboard. The coding and analysis can then be parallelized.

5.6 Takeaway

In this chapter, we have discussed several techniques to solve NLP problems. For "simple" problems, direct substitution might work. The Lagrange multiplier method can be used if information concerning the derivatives is available. Penalty and barrier methods can be used to take constraints into the objective function (sometimes called Lagrangian relaxation). Quadratic programming is based on reformulating a quadratic program into a linear program. SLP linearizes nonlinear systems and solves the resulting LP subproblems iteratively. SQP approaches non-linear problems as quadratic problems and quadratic problems can be solved with linear engines. NLP can be encountered in reactor design (maximize the yield, while encountering nonlinear kinetics), process flowsheeting (process simulators for steady-state simulation), and parameter estimation (fit nonlinear models to VLE data).

Further Reading

Bazaraa, M. S., Sherali, H. D., & Shetty, C. M. (2006). *Nonlinear programming: Theory and algorithms* (3rd ed.). Wiley.

Edgar, T. F., Himmelblau, D. M., & Lasdon, L. S. (2001). *Optimization of chemical processes* (2nd ed.). McGraw-Hill.

Finlayson, B. A. (2011). *Numerical methods for chemical engineers with MATLAB applications* (2nd ed.). Prentice Hall.

Biegler, L. T., & Zavala, V. M. (2009). Recent advances in successive quadratic programming. *Computers & Chemical Engineering*, *33**(3), 575–582.

NEOS Guide. (n.d.). *Nonlinear optimization*. https://neos-guide.org

Luenberger, D. G. (1984). *Applied nonlinear programming*. Addison-Wesley.

6 Integer and Mixed-Integer Programming

When decisions come in whole numbers.

6.1 Introduction to Discrete Optimization

Many chemical engineering problems, for example, in operations, design, location, and scheduling, involve discrete decisions. We distinguish between so-called binary decisions, such as *yes/no decisions*, to install or not install a new piece of equipment. Sometimes binary decisions are also called 0–1 decisions. Besides binary decisions, we have also *integer decisions*, i.e., real numbers 0,1,2,3, . . . You can think, for example, of the number of stages in a distillation column. The class of optimization problems that involve integer variables are called *integer programs* (IPs) or *mixed-integer programs* (MIPs), if there are discrete as well as continuous decision variables present. Many MIPs are linear in the objective function and constraints, and for that reason, they are subject to solution by linear programming (LP). These problems are called *mixed-integer linear programs* (MILPs). Problems where the objective and/or constraints are nonlinear are called *mixed-integer nonlinear programs* (MINLPs). We can find MIPs in many application areas, in process synthesis, in equipment selection, and in scheduling activities. The key differences between LPs and NLPs are that MIPs are NP-hard problems (with large combinatorial complexity) and that Simplex method (for LPs) and gradient methods (for NLPs) are replaced or extended with branch-and-bound algorithms.

6.2 Problem Formulation

We might organize IPs in three main classes, which have been discussed in the operations research literature: 1) the knapsack problem, 2) the traveling salesman problem, and 3) the blending problem.

The Knapsack Problem: In the knapsack problem, we have n objects. The weight of the ith object is w_i, and its value is v_i. Select a subset of the objects such that their total weight does not exceed W (the capacity of the knapsack), and their total value is a maximum:

$$\max f(y) = \sum_i^n v_j y_i$$

subject to: $\sum_i^n w_i y_i \leq W$

$$y_i \in \{0,1\}$$

https://doi.org/10.1515/9783111342283-006

The binary variable y_I indicates whether an object I is selected (($y_i = 1$) or not selected ($y_i = 0$). The knapsack problem in its different forms can, for example, be encountered in crude inventory management at refineries.

The Blending Problem: A list of possible ingredients is provided. The ingredients should be blended into a product, from a list containing the weight, value, cost and analysis of each ingredient. The objective is to select from the list a set of ingredients so as to have a satisfactory total weight and analysis at minimum cost for a blend. Let x_j be the quantity of ingredient j be available in the continuous amounts and let y_k represent the ingredients to be used in discrete quantities v_k. So $y_k = 1$ if the ingredient is used, and if not, let c_j and d_k be the respective costs of the ingredients and be the fraction of component i in ingredients j:

$$\min f(x,y) = \sum_j c_j x_j + \sum_k d_k v_k y_k$$

subject to: $W^l \leq \sum_j x_j + \sum_k v_k y_k \leq W^u$

$$A_i^l \leq \sum_j a_{ij} x_j + \sum_k a_{ik} v_k y_k \leq A_i^u$$

$$0 \leq x_j \leq u_j \ \forall j$$

$$y_k \in \{0,1\} \ \forall k$$

where u_j is the upper limit of the jth ingredient, W^l and W^u are the lower and upper bounds on the weights, and A_i^l and A_i^u are the lower and upper bounds on the analysis for component i.

The Traveling Salesman Problem: The problem is to assign values of 0 or 1 to the variables y_{ij} where $y_{ij} = 1$ if the salesman travels from city I to city j and $y_{ij} = 0$ otherwise. The constraints in the problem are that the salesman must start at a particular city, visit each of the other cities only once, and return to the original city. A cost c_{ij} is associated with traveling from city I to city j (often the distance between the two places). The objective is to minimize the total cost of the trips to each city visited:

$$\min f(y) = \sum_i^n \sum_j^n c_{ij} y_{ij}$$

subject to:

$$\sum_i^n y_{ij} = 1 \ \forall j$$

$$\sum_j^n y_{ij} = 1 \ \forall j$$

$$i, j = 1, \ldots, n$$

$$y_{ii} = 0$$

$$y_{ij} \in \{0,1\}$$

The two equality constraints (often called assignment constraints) ensure that each city is only visited once in any direction. $y_{ii} = 0$ because no trip is involved.

For a chemical plant analogy, the problem can also be cast in terms of processing n batches on a single piece of equipment, in which the equipment is reset between processing the ith and jth batches. The batches can be processed in any order. The cost is in this case the time or cost to set up the equipment to do batch j if it was previously doing batch I, and $y_{ij} = 1$ means that batch I is immediately followed by batch j.

In a same way, the traveling salesman problem can also be used to formulate superstructure optimization problems, where the goal is to find cost-effective routes from raw material to final product via selecting the appropriate technologies in-between.

6.3 Solution Methods

There are several methods around that can be used to solve MIPs, all with their specific strengths and weaknesses.

6.3.1 The *Outer Approximation* Method

Outer approximation (OA) is an iterative method for solving MINLPs by alternating between solving: 1) nonlinear subproblems (fixed integer variables) to provide primal bounds and and 2) MILP master problems (linearizations of nonlinear constraints) to generate cuts and improve the solution. At each iteration, the master problem refines a polyhedral OA of the feasible region, converging to the global optimum for convex problems. OA is widely used in process optimization (e.g., reactor design and flowsheet synthesis) due to its efficiency in handling nonconvexities via tailored cutting planes (see Section 6.3.4).

6.3.2 *Benders Decomposition*

Benders decomposition is a divide-and-conquer algorithm for solving large-scale optimization problems (e.g., MILPs or stochastic programs) by splitting them into the following: 1) a master problem (handling integer/complicating variables) and 2) a subproblem (solving the remaining linear/continuous variables for fixed integers). The master problem proposes trial integer solutions, while the subproblem generates

Benders cuts (feasibility or optimality constraints) to iteratively tighten the master's approximation. This method shines in stochastic programming and process design, where problems decompose naturally (e.g., design-operations splits).

6.3.3 Branch and Bound

The *branch-and-bound method* is the basis for many algorithms used for solving MIPs. It effectively relaxes the original problem without integer constraints and then splits up the problem into subproblems (branching) while fixing some of the variables (bound). It will prune the subproblems worse than the current best.

Example: Let us now consider the following IP:

$$\max z = 8x_1 + 11x_2 + 6x_3 + 4x_4$$

subject to:

$$5x_1 + 7x_2 + 4x_3 + 3x_4 \leq 14$$

$$x_j \in \{0,1\} \ \forall j$$

The variable x_j is either 0 or 1, as reflected in the *integrality constraint*. We could actually try all possible combinations of 0 and 1 for all x's. We would have to check 16 possible combinations, as shown in Table 6.1.

Table 6.1: All possible solutions to our example.

Solution no.	X_1	X_2	X3	X_4	Z	RHS
1	0	0	0	0	0	0
2	0	0	0	1	4	3
3	0	0	1	0	6	4
4	0	0	1	1	10	7
5	0	1	0	0	11	7
6	0	1	0	1	15	10
7	0	1	1	0	17	11
8	0	1	1	1	21	14
9	1	0	0	0	8	5
10	1	0	0	1	12	8
11	1	0	1	0	14	9
12	1	0	1	1	18	12
13	1	1	0	0	19	12
14	1	1	0	1	23	15
15	1	1	1	0	25	16
16	1	1	1	1	29	19

As we can see from Table 6.1, solutions 14, 15, and 16 are infeasible, the right-hand-side (RHS) of the constraint exceeds the value of 14. From the remaining solutions, Solution #8 has the highest objective value of $z = 14$.

You can imagine that if the number of variables and constraints increases, we have to perform many more calculations. The combinatorial complexity goes up with the number of variables. In addition, we are also evaluating infeasible solutions when we solve the problem with a direct approach as described above.

The branch-and-bound algorithm
1) relaxes the integrality constraint,
2) solves the resulting LP, and then
3) creates two subproblems for non-integer solutions.

So our example now becomes an LP which can be solved with the Simplex method:

$$\max z = 8x_1 + 11x_2 + 6x_3 + 4x_4$$

subject to:

$$5x_1 + 7x_2 + 4x_3 + 3x_4 \leq 14$$

$$0 \leq x_j \leq 1 \; \forall j$$

We will find that $x_1 = 1$, $x_2 = 1$, $x_3 = 0.5$, and $x_4 = 0$. The objective value is $z = 22$. We call the objective value of this first solution the *upper bound solution*. It can also be noticed that only x_3 has not taken an integer value. We will now formulate two new problems (subproblems) where we fix to a value of 0 in one case and to 1 in the other case.

When we fix $x_3 = 0$ and solve the LP, we find that $x_1 = 1, x_2 = 1, x_4 = 0.66, z = 21.65$.

When we fix $x_3 = 1$ and solve the LP, we find that $x_1 = 1, x_2 = 0.714, x_4 = 0, z = 21.85$.

We could visualize the original LP and the two subproblems that branch on x_3 in Figure 6.1.

When fixing $x_3 = 0$ that x_4 is not an integer. And when we fix $x_3 = 1$, we see that x_2 is not an integer. We typically start branching in the part of the tree where the objective function has the highest value.

So in one subproblem, we fix $x_2 = 0$ mind that we already fixed $x_3 = 1$. We now find that $x_1 = 1, x_4 = 1$ and $z = 18$. This is actually an integer solution. We cannot further branch in this line of the tree, and the tree is *fathomed*. We might stop, but there might be a better solution in the other branch.

In the second subproblem, we fix $x_2 = 1$, and we already fixed $x_3 = 1$. Here we find that $x_1 = 0.6, x_4 = 0$ and $z = 21.8$.

This second set of subproblems can be visualized in Figure 6.2.

Now x_1 is not an integer, we can create two new subproblems where we fix x_1 to either 1 or 0.

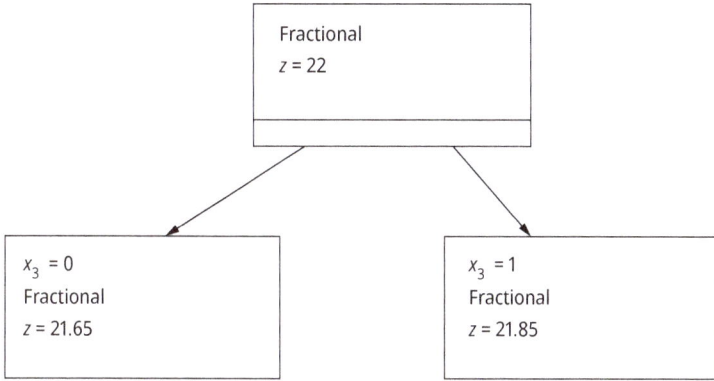

Figure 6.1: The upper bound solution and two subproblems branched on x_3.

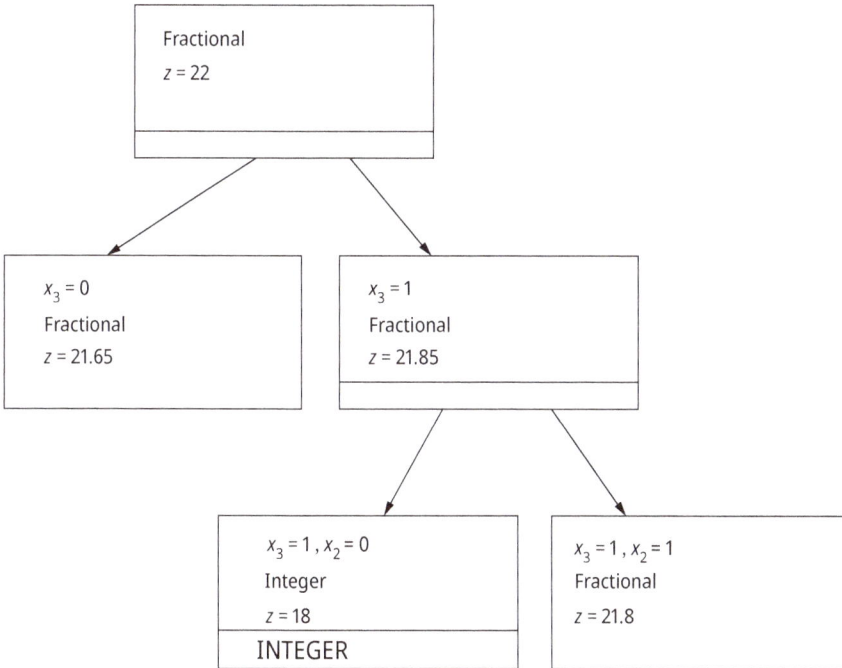

Figure 6.2: Two new subproblems branched on x_2.

In the first subproblem, we fix $x_1 = 0$ mind that also $x_2 = 1, x_3 = 1$ are fixed. We will find that $x_4 = 1, z = 21$. This is again an integer solution.

In the second subproblem, we fix $x_1 = 1$ mind that also $x_2 = 1, x_3 = 1$ are fixed. This leads to an infeasible problem, and the constraint is no longer satisfied.

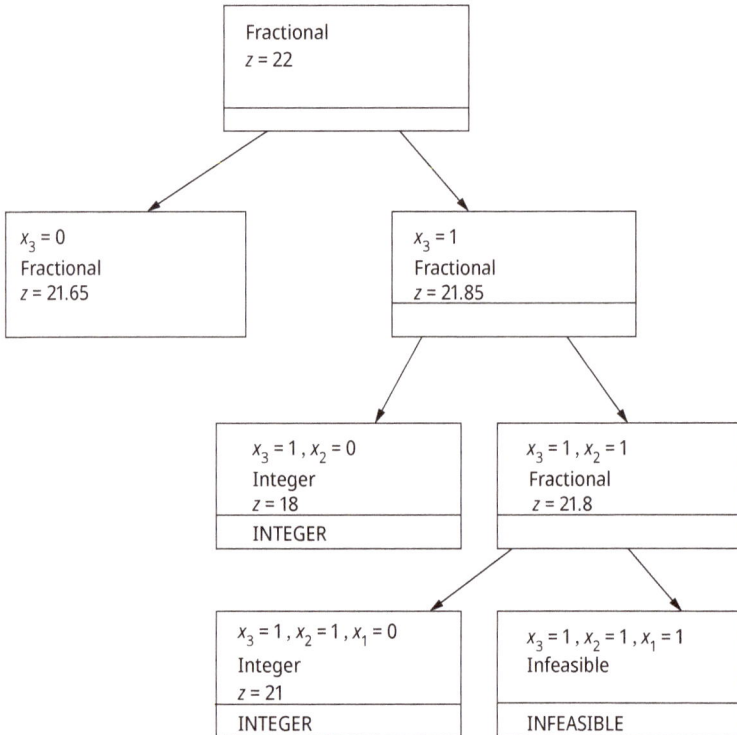

Figure 6.3: Final branch on x1.

6.3.4 Cutting Planes

The *cutting planes method* is an iterative algorithm for solving MIPs by progressively refining the feasible region with linear inequalities (cuts) that exclude fractional solutions while preserving integer optima. These cuts are also called *Gomory cuts*.

Let us see how a Gomory cut can be created with an example. Suppose we have the following integer LP:

$$\max Z = 5x_1 + 8x_2$$

subject to:

$$x_1 + x_2 \leq 6$$

$$5x_1 + 9x_2 \leq 45$$

$$x_1, x_2 \in \{0, 1, 2, 3, \ldots\}$$

Step 1: Solve the LP *Relaxation*

We relax the integrality constraint and solve a regular LP; if the outcome is fractional, we will have to create a Gomory cut.

When we solve the relaxed LP, we will find $x_1 = 4.5$, $x_2 = 3.5$, and $Z = 63$.

Step 2: Identify the Source Row

We have to look into the Simplex table and pick the row where a basic variable is fractional. We can use the row for x_1:

$$x_1 + 0.2s_1 - 0.1s_2 = 4.5$$

s_1 and s_2 are the slack variables.

Step 3: Separate Fractional Parts

For each coefficient in the row, split into integer and fractional parts:

$$x_1\!:\!1 = 1 + 0$$

$$s_1\!:\!0.2 = 0 + 0.2$$

$$s_2\!:\!-0.1 = -1 + 0.9$$

$$RHS\!:\!4.5 = 4 + 0.5$$

Step 4: Formulate the Gomory Cut

Construct the cut using only the fractional parts:

$$0.2s_1 + 0.9s_2 \geq 0.5$$

Note: The sum of the fractional coefficients must be larger than or equal to the fractional part of the RHS to force an integer solution.

Step 5: Convert Back to Original Values

Express the slack variables in terms of x_1 and x_2:

$$s_1 = 6 + x_1 - 3x_2$$

$$s_2 = 35 - 7x_1 - x_2$$

Substitute in the cut and rearrange:

$$6.1x_1 + 1.5x_2 \geq 31.7$$

This cut can be added as a new constraint to the LP. The new solution will "cut away" the fractional solution, moving toward an integer optimum.

From the Expert: **Prof. Ignacio Grossmann**
Carnegie Mellon University, Pittsburgh, USA

Evolution of Optimization in Process Systems Engineering (PSE)

Optimization models and algorithm have played a major role in PSE through its application to many relevant problems in the chemical process industry. LP found one of its first applications in the optimization of production planning of petroleum refineries, specifically by Charnes and Cooper [4], professors at Carnegie Institute of Technology, who collaborated with Gulf Oil in Pittsburgh. Also, within the oil industry, Martin Beale and Small [2] from BP developed one of the first codes for solving MILP models. Modern codes for MILP such as Gurobi, CPLEX, and XPRESS are currently used not only by oil companies such as ExxonMobil and Shell, but also by major chemical manufacturing companies such as Dow Chemical and BASF, and by industrial gases such as Air Liquide, Air Products and Lindt. These MILP models have been aimed mostly at the planning and scheduling of industrial supply chains.

Motivated by the nonlinear optimization models that arise in chemical processes, this has led to the development of nonlinear programming (NLP) algorithms and codes by researchers in PSE. While the original theory of NLP was developed by researchers in Operations Research and Applied Mathematics [1], the state-of-the-art code IPOPT for NLP was developed in chemical engineering at Carnegie Mellon by Waechter and Biegler [8]. Going a step further, to address the optimization of discrete variables through MINLP problems, as for instance in superstructure optimization problems, Duran and Grossmann [5], from chemical engineering at Carnegie Mellon, developed the outer-approximation algorithm that is currently implemented in the software DICOPT. To address the issue that often NLP and MINLP problems are nonconvex, possibly giving rise to local optima, global optimization algorithms have been developed, first by Misener and Floudas [9], chemical engineering at Princeton, and subsequently by the chemical engineers Tawarmalani and Sahinidis [10] who produced the state-of-the-art global optimizer code BARON. Researchers in PSE have also been active in the area of stochastic programming to address the optimization under uncertainty [8].

From the above, it is clear that there have been many successful optimization methods and codes developed by researchers in PSE, and successfully applied in many PSE applications (e.g., see [3, 7]).

References

[1]   Bazaraa, Mokhtar S. and Shetty, C. M. (1979). Nonlinear programming. Theory and algorithms. John Wiley & Sons. ISBN 0-471-78610-1.

[2]   Beale, E. M. L., R. E. Small. 1965. MIP by a branch and bound technique, Proc. IFIP Congress, Vol. 2 (W. Kalench, Ed.), Macmillan, London (1965) 450–451.

[3]   Biegler, L.T., "New nonlinear programming paradigms for the future of process optimization," *AIChE J.*, 63, 178–1193 (2017).

[4]   Chen, Q., E.S. Johnson, D.E. Bernal, R. Valentin, S.Kale, J. Bates, J. D. Siirola and I.E. Grossmann, "Pyomo.GDP: an ecosystem for logic based modeling and optimization development," *Optimization and Engineering* 23, 607–642 (2022)

[5]   Cooper, W.W., "Abraham Charnes and W. W. Cooper (et al.): a brief history of a long collaboration in developing industrial uses of linear programming," Operations Research 50(1):35–41 (2002).

[6]   Duran, M.A. and I.E. Grossmann, "An Outer-Approximation Algorithm for a Class of Mixed-integer Nonlinear Programs," *Math Programming* 36, 307 (1986).

[7]   Grossmann, I.E., "Advances in Mathematical Programming Models for Enterprise-Wide Optimization," *Computers & Chemical Engineering*, 47, 2–18 (2012).

[8]   Li, C. and I.E. Grossmann, "A Review of Stochastic Programming Methods for Optimization of Process Systems under Uncertainty," *Frontiers of Chemical Engineering* 2,622241(2021).

[9]   Misener, R., CA Floudas, "ANTIGONE: algorithms for continuous/integer global optimization of nonlinear equations," Journal of Global Optimization 59 (2), 503–526

[10] Tawarmalani, M.; Sahinidis, N.: Global Optimization of Mixed Integer Nonlinear Programs: A Theoretical and Computational Study. *Mathematical Programming*, 99 (3), 563–591 (2004).

[11] Waechter, A. and L. T. Biegler, "On the implementation of an interior-point filter line-search algorithm for large-scale nonlinear programming," Math. Program., 106, 25–57 (2006)

In Figure 6.4, the Gomory cut is visualized. It also shows the LP feasible region (shaded), the integer solutions (black dots), the fractional optimum (red dot), and the Gomory cut (dashed line).

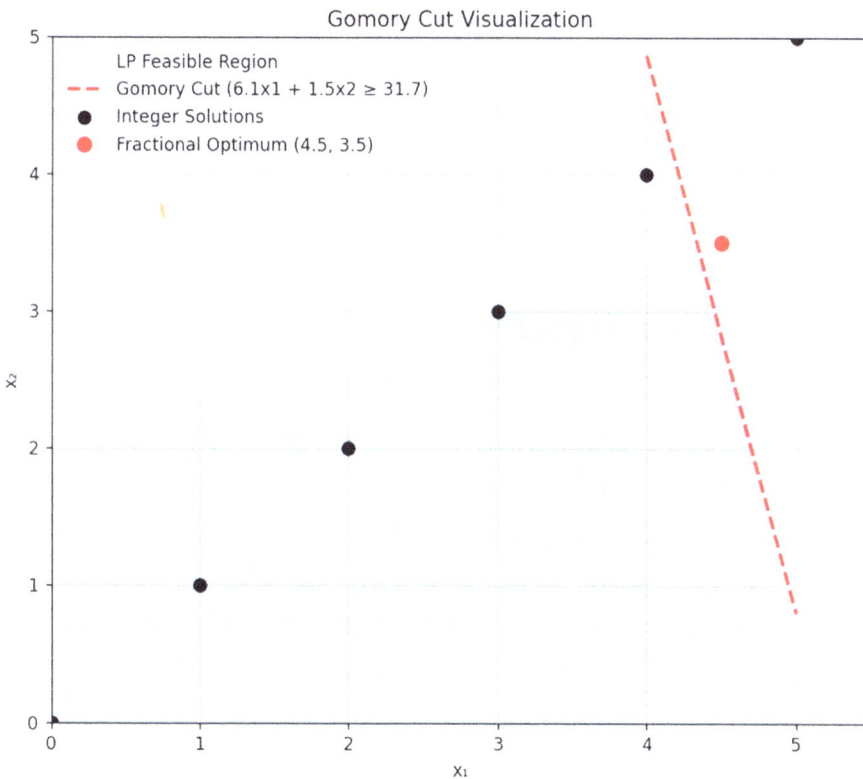

Figure 6.4: Gomory cut visualization.

By adding Gomory cuts, we can preserve all integer solutions. The Gomory cut will tighten the feasible region by excluding fractional solutions. Although by cutting planes, there is guaranteed convergence, it might be slow.

Here is another example that implements the cutting planes method. Consider the following problem:

$$\max Z = 5x_1 + 8x_2$$

subject to:

$$x_1 + x_2 \leq 6$$

$$5x_1 + 9x_2 \leq 45$$

$$x_1, x_2, \geq 0$$

$$x_1, x_2 \in \{0, 1, 2, 3, \ldots\}$$

Step 1: Solve the LP Relaxation
We will find the optimal solution to be $x_1 = 2.25, x_2 = 3.75, Z = 41.25$ Both x_1 and x_2 are non-integers.

Step 2: Generate the Gomory Cut
From the second constraint in the Simplex table:

$$x_2 + 1.25s_1 - 0.25s_2 = 3.75$$

where s_1 and s_2 are the slack variables, and we generate the Gomory cut:

$$0.25s_1 + 0.75s_2 \geq 0.75$$

Convert the slack variables back to x_1, x_2:

$$x_1 + x_2 \leq 6 \rightarrow s_1 = 6 - x_1 - x_2$$

$$5x_1 + 9x_2 \leq 45 \rightarrow s_2 = 45 - 5x_1 - 9x_2$$

Substitute in the cut and rearrange:

$$4x_1 + 7x_2 \geq 39$$

Step 3: Resolve the LP with the Cut
We will add the cut as a new constraint and find after solving: $x_1 = 1.8, x_2 = 4, Z = 41$. We see that x_1 is still fractional.

Step 4: Add a Second Cut
From the new Simplex table row for x_1:

$$x_1 - 1.4s_1 + 0.4s_3 = 1.8$$

s_3 is the slack for the new cut. The Gomory cut is:

$$0.6s_1 + 0.4s_3 \geq 0.8$$

Convert to x_1, x_2 and substitute:

$$x_1 + x_2 \leq 5$$

Step 5: Final LP Solution
Add the Gomory cut to the LP and resolve. We will now find $x_1 = 0, x_2 = 5, Z = 40$. This is an integer optimal solution.

In three calculation steps, we have found the optimum, as summarized in Table 6.2.

Table 6.2: Summary of the cutting plane method for our example.

Iteration	Solution	Z	Action
1 (Initial)	$x1 = 2.25, x_2 = 3.75$	41.25	Add cut $4x_1 + 7x_2 \geq 39$
2	$x1 = 1.8, x_2 = 4$	41	Add cut $x_1 + x_2 \leq 5$
3 (Final)	$x1 = 0, x_2 = 5$	40	**Integer optimal**

6.4 Linearization

We have so far discussed IPs, which could be converted to linear programs. For MILPs, this conversion to LPs is very similar. However, when we are dealing with MINLPs, we might run into problems related to convexity and continuity. One common practice is to bypass nonlinearity by converting the constraints or objective to linear constraints.

6.4.1 The *Glover Linearization*

Glover linearization can be used to convert certain nonlinear terms in an optimization problem into an equivalent linear form, making them solvable with LP or MILP techniques. The Glover linearization is particularly useful when dealing with products of binary and continuous variables.

We can encounter such nonlinearities in many chemical engineering problems, for example, in process selection (choosing between different reactors with varying costs), in batch scheduling (deciding whether to run a unit a certain time) or in fixed cost problems (activating a distillation column will incur costs only when used).

Consider a term $z = yx$ where $y \in \{0,1\}$ is a binary variable and x is a continuous variable that is bounded between lower and upper bounds $L \leq x \leq U$.

We introduce a new variable z and enforce the following constraints:

$$z \leq Uy$$

$$z \geq Ly$$

$$z \leq x - L(1-y)$$

$$z \geq x - U(1-y).$$

These constraints ensure that if $y = 1$, then $z = x$, and that if $y = 0$ then $z = 0$.

Example: Suppose (trivial) a chemical plant must decide whether to use a batch reactor $(y = 1)$ or not $(y = 0)$. The reactor has a fixed cost of \$500 (if used) and a variable cost of \$2 per kilogram of product. The production rate x must be between 0 and 200 kg. We could write this as follows:

$$\min Z = 500y + 2xy$$

subject to: $0 \leq x \leq 200$

This is an MINLP because the objective function is nonlinear. We can now apply the Glover reformulation to convert the problem into an MILP:

$$\min Z = 500y - 2z$$

subject to:

$$z \leq 200y$$

$$z \geq 0$$

$$z \leq x$$

$$z \geq x - 200(1-y)$$

If $y = 1$ then $z = x$. (the reactor is used; costs depend on production), and if $y = 0$ then $z = 0$ (the reactor is not used, so the costs are zero). The result is an MILP that can be solved with linear solvers. In case the upper bound is not defined, we can use an arbitrarily large number. This number is often referred to as the *big-M*. Figure 6.5 shows how PhD student Prya is using the big-M to solve a biorefinery optimization problem.

We could go to a simpler formulation where we get rid of the auxiliary variable z. We can replace xy with x and add the following two constraints:

$$Ly \leq x$$

$$x \le Uy$$

We will achieve the same if $y = 1$ then $L \le x \le U$, and if $y = 1$ then $x = 0$. This shortcut linearization is called the standard linearization.

Let us consider a simple production planning problem, formulated as an MILP: A factory produces two products (A and B) with a profit of \$30 per unit of A and \$30 per unit of B. The machine time differs per type of product, for product A: 3 h/unit for machine 1 and 2 h per unit for machine 2. For product B: 2.5 h per unit for machine 1 and 4 h per unit for machine 2.

The main constraint is in the capacity, machine 1 has 100 h available, and machine 2 has 120 h available.

The discrete decision is that product A requires a setup cost of \$100 if any units are produced. We can formulate this problem as an MILP:

$$\max Z = 30x_A + 30x_B - 100y_A$$

subject to:

$$3x_A + 2x_B \le 100 \, (\text{machine 1})$$

$$2.5x_A + 4x_B \le 120 \, (\text{machine 2})$$

$$x_A \le My_A \, (\text{Setup cost})$$

$$x_A, x_B \ge 0, y_A \in \{0,1\}$$

where M is the *big-M*, for example 100.

To find the solution to this problem, we could work out a Python implementation, using the PuLP library:

```
from pulp import *

# Initialize model
model = LpProblem("Production_Planning", LpMaximize)

# Variables
x_A = LpVariable("x_A", lowBound=0, cat='Continuous') # Units of A
x_B = LpVariable("x_B", lowBound=0, cat='Continuous') # Units of B
y_A = LpVariable("y_A", cat='Binary')                 # Setup for A

# Objective: Maximize profit
model += 30*x_A + 30*x_B - 100*y_A

# Constraints
model += 3*x_A + 2*x_B <= 100, "Machine1"
```

```
model += 2.5*x_A + 4*x_B <= 120, "Machine2"
model += x_A <= 100*y_A, "Setup_Cost"
# If x_A > 0, y_A must be 1
# Solve
model.solve()

# Results
print(f"Status: {LpStatus[model.status]}")
print(f"Produce {x_A.varValue:.0f} units of A (Setup: {y_A.
varValue})")
print(f"Produce {x_B.varValue:.0f} units of B")
print(f"Total Profit: ${value(model.objective):.2f}")
```

After running this script, the output will be as follows:

```
Produce 23 units of A (Setup: 1.0)
Produce 16 units of B
Total Profit: $1057.14
```

6.4.2 The *McCormick Envelope*

The *McCormick envelope* is a mathematical technique used to linearize the product of two continuous variables ($z = x \cdot yz = x \cdot y$) in optimization problems by constructing the tightest possible convex relaxation. It replaces the nonlinear bilinear term with **four linear inequalities** derived from the bounds of xx and yy, creating a polyhedral envelope that contains all feasible solutions. These constraints ensure zz approximates $x \cdot yx \cdot y$ while maintaining convexity, enabling the use of efficient linear or mixed-integer solvers. The envelope is exact at the variable bounds (e.g., when xx and yy take their minimum or maximum values) and provides valid over- and underestimators elsewhere. Widely used in chemical engineering for problems such as reactor design or heat exchanger optimization, McCormick envelopes balance accuracy with computational tractability, though their tightness depends heavily on the quality of the variable bounds. For higher-order nonlinearities, they are often combined with other techniques (e.g., piecewise linearization).

6.4.3 The *Lambda Method*

The purpose of the *lambda method* is to approximate a nonlinear function $f(x)$ using linear segments (breakpoints). The key idea is that x and $f(x)$ are represented as weighted sums of breakpoint values, where the weights λ_i are the optimization variables.

Figure 6.5: How PhD student Priya uses a big-M implementation to converge an MILP.

First we have to define breakpoints (x_i, f_i) for $i = 1, \ldots, n$. Then, we express x and $f(x)$ as convex combinations:

$$x = \sum_i^n \lambda_i x_i$$

$$f(x) \approx \sum_i^n \lambda_i f_i$$

We have to ensure convexity:

$$\sum_i^n \lambda_i = 1$$

Non-negativity:

$$\lambda_i \geq 1$$

And make sure that only two adjacent points λ_i can be nonzero (this is also called a special ordered set of Type 2, SOS2).

Let us consider as an example the piecewise linearization of $f(x) = x^2$ over the interval $x \in [0,4]$. We will use breakpoints at $x = [0,1,2,3,4]$, giving the following equation:

x_i	f_i
0	0
1	1
2	4
3	9
4	16

The key idea is that between two adjacent points (x_i, f_i) and (x_{i+1}, f_{i+1}), we approximate $f(x)$ as a straight line.

For any x in $[x_i, x_{i+1}]$, we express x and $f(x)$ as the weighted sum of the two nearest breakpoints using variables λ_i:

$$x = \sum_i^4 \lambda_i x_i$$

$$f(x) \approx \sum_i^4 \lambda_i f_i$$

$$\sum_i^4 \lambda_i = 1$$

$$\lambda_i \geq 1$$

For example, if $x = 1.5$, then x lies between $x_1 = 1$ and $x_2 = 2$, the weights are then $\lambda_1 = 0.5$.

And $\lambda_2 = 0.5$. Then $f(x) \approx 0.5 \times 1 + 0.5 \times 4 = 2.5$

Say, if $x = 3.75$, then x lies between $x_3 = 3$ and $x_4 = 4$, the weights are then $\lambda_3 = 0.25$. And $\lambda_4 = 0.75$. Then $f(x) \approx 0.25 \times 9 + 0.75 \times 16 = 14.25$

Below, you can find a MATLAB code that can determine the weights for our example. Here we are calling the solve statement and specifically addressing the intlinprog algorithm:

```matlab
% Piecewise linear approximation of f(x) = x^2 at x = 1.5 using SOS2
% method
% Define breakpoints
breakpoints = [0, 1, 2, 3, 4];
f_values = breakpoints.^2;
n = length(breakpoints);
target_x = 1.5; % The x-value we want to evaluate
% Create optimization model
model = optimproblem('Description','Piecewise Linear Approximation
at x=1.5');
% Variables
lambda = optimvar('lambda', n, 'LowerBound', 0);
f = optimvar('f');
% Constraints
model.Constraints.convexity = sum(lambda) == 1;
model.Constraints.x_def = target_x == sum(lambda.*breakpoints'); %
Fixed x value
model.Constraints.f_def = f == sum(lambda.*f_values');

% SOS2 constraint implementation
z = optimvar('z', n-1, 'Type', 'integer', 'LowerBound', 0,
'UpperBound', 1);
model.Constraints.sos1 = lambda(1) <= z(1);
for i = 2:n-1
   model.Constraints.(['sos' num2str(i)]) = lambda(i) <= z(i-1) + z(i);
end
model.Constraints.(['sos' num2str(n)]) = lambda(n) <= z(n-1);
model.Constraints.one_active = sum(z) == 1;

% Objective: Just evaluate f (no optimization needed, but we keep the
structure)
model.Objective = f;

% Solve
options = optimoptions('intlinprog', 'Display', 'off');
[sol, fval, exitflag] = solve(model, 'Options', options);

% Results
if exitflag > 0
   fprintf('Evaluation at x = %.2f\n', target_x);
   fprintf('Approximated f(%.2f) = %.4f\n', target_x, sol.f);
```

```
    fprintf('True f(%.2f) = %.4f\n', target_x, target_x^2);
    fprintf('Error = %.4f\n\n', abs(sol.f - target_x^2));


    fprintf('Breakpoint weights (lambda values):\n');
    for i = 1:n
        fprintf(' λ _%d (x=%.1f) = %.4f\n', i-1, breakpoints(i), sol.
lambda(i));
    end


    % Verify the weighted sum
    fprintf('\nVerification:\n');
    fprintf('Sum of lambdas: %.4f\n', sum(sol.lambda));
    fprintf('Weighted x sum: %.4f (should equal %.4f)\n', ...
            sum(sol.lambda.*breakpoints'), target_x);
    fprintf('Weighted f(x) sum: %.4f\n', sum(sol.lambda.*f_values'));
else
    error('Solver failed to find solution');
end
```

The integer variables in this MATLAB implementation are used to enforce the SOS2 constraint. MATLAB's intlinprog does not support SOS2 constraints (as in other packages), so it has to be modeled manually using binary variables.

There will be a binary variable for each active segment between breakpoints. Each z_i corresponds to the interval $[x_i, x_{i+1}]$. In other words, if $z_i = 1$, the solution lies in $[x_i, x_{i+1}]$. Only one z_i can be one.

The segment activation can now be enforced by the following equation:

$$\lambda_1 \leq z_1$$

$$\lambda_i \leq z_{i+1} + z_i, \text{ for } i = 2, \ldots, n-1$$

$$\lambda_n \leq z_{n-1}$$

And to make sure that there is only one active segment, we need the following equation:

$$\sum_i^n z_i = 1$$

In other words, the binary variable acts as a switch for each segment.

In Figure 6.6, you can see the piecewise linearization of $f(x) = x^2$.

The lambda method is extremely useful to keep your models linear. We often encounter nonlinearity in setting up chemical engineering models. You can think, for example, of nonlinear relationships for costs versus capacity (economies of scale) or reaction kinetics or thermodynamic expressions.

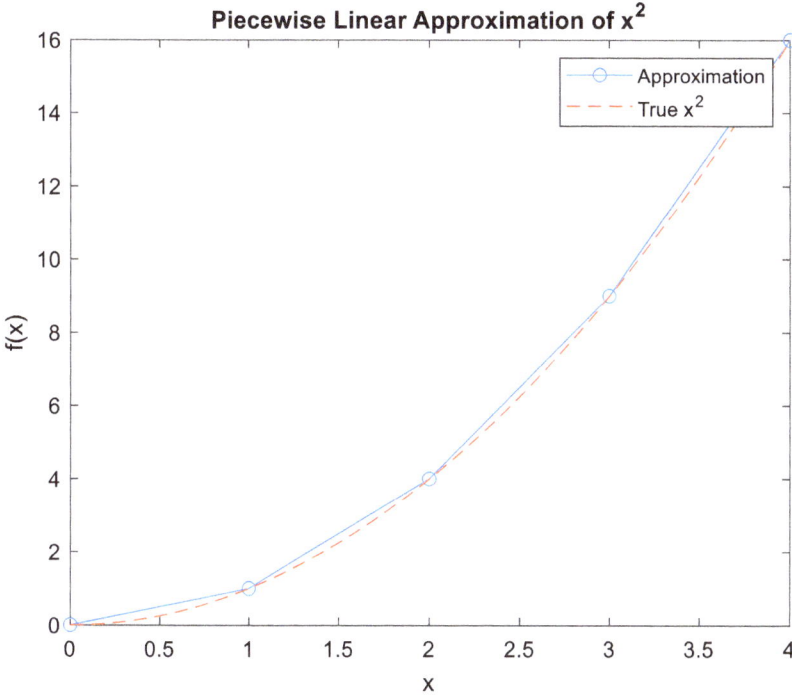

Figure 6.6: Piecewise linear approximation of a parabola.

6.5 *Logic Inference*

Logic inference is the systematic process of encoding "if-then," "either-or," and other logical relationships between decisions using binary variables and constraints. In chemical engineering optimization, this allows modeling complex operational rules and safety requirements.

The core concepts are as follows: 1) Binary variables are used to represent true/false decisions (e.g., $y = 1$ if a reactor is used, and else $y = 0$); and 2) the use of logical operators AND ($A \cap B$), OR ($A \cup B$), Implication ($A \Rightarrow B$ if A then B), and negation ($\neg A$, not A).

Here are a few examples:

Example 1 (Implication): If a feedstock is selected ($F1$), then the storage tank must be heated ($T1$):

$$T1 \geq F1$$

If $F1 = 1$, it forces $T1 = 1$.

Example 2 (Disjunction) (Either-Or): Either use pipeline A or pipeline B, but not both:

$$y_A + y_B \leq 1$$

Example 3 (Conditional Constraints): If the reactor operates above 200 degrees ($y = 1$), then cooling must be active ($z = 1$):

$$z \leq y,$$

$$\text{Cost}_{\text{cooling}} = 500z$$

There are structured procedures available to derive such *logical constraints*. Let us go through such a procedure via an example.

We want to translate the following *heuristic* into a logical constraint:

> "If the absorber is selected to recover the product or the membrane, then do not use the cryogenic separator."

First, we assign Boolean literals to each action P_A = select absorber, P_M = select the membrane separator, P_C = select cryogenic separation. The logic expression is given by the following equation:

$$P_A \cup P_M \Rightarrow \neg P_C$$

We can remove the implication ($P_1 \Rightarrow P_2 \Leftrightarrow \neg P_1 \cup P_2$):

$$\neg(P_A \cup P_M) \cup \neg P_C$$

Apply De Morgan's theorem ($\neg(P_1 \cap P_2) \Leftrightarrow \neg P_1 \cup \neg P_2$ and $\neg(P_1 \cup P_2) \Leftrightarrow \neg P_1 \cap \neg P_2$):

$$(\neg P_A \cap \neg P_M) \cup \neg P_C$$

Distributing the OR over the AND gives the following:

$$(\neg P_A \cup \neg P_C) \cap (\neg P_M \cup \neg P_C)$$

Assigning the corresponding 0–1 variables to each term in the above *conjunction* leads to the following:

$$(1 - y_A) + (1 - y_C) \geq 1$$

$$(1 - y_M) + (1 - y_C) \geq 1$$

which can be rearranged to the following two inequalities:

$$y_A + y_C \leq 1$$

$$y_M + y_C \leq 1$$

6.6 Computational Tools

Gurobi excels in solving large-scale MIP problems with its advanced algorithms and parallel processing, often achieving optimal solutions faster than competitors, especially for complex process scheduling and network design problems. However, its commercial licensing can be prohibitively expensive for small organizations. **CPLEX** similarly offers robust performance and excellent support for nonlinearities through piecewise linearization, making it ideal for refinery optimization, but its steep learning curve and high memory usage may challenge novice users. **COIN-OR CBC**, as an open-source alternative, provides decent performance for medium-sized problems like equipment selection or blending optimization, though it lacks the cutting-edge heuristics and presolving capabilities of commercial solvers, sometimes struggling with tightly constrained problems. **SCIP** stands out for its flexibility in handling complex logical constraints and non-convex MINLPs, useful for safety system design, but its slower solution times make it less suitable for real-time applications. **Google OR-Tools** is beginner-friendly with Python integration, perfect for educational purposes or small-scale production planning, but its limited advanced features reduce effectiveness for industrial-scale problems. For chemical engineers, the choice hinges on problem scale, budget, and need for advanced features – commercial solvers (Gurobi/CPLEX) dominate performance-critical applications, while open-source tools (CBC/SCIP) suffice for prototyping or constrained budgets.

6.7 Challenges and Workarounds

Solving MIP problems in chemical engineering presents several fundamental challenges that require careful consideration and strategic approaches to overcome.

One of the most pervasive issues is the curse of dimensionality, where large-scale problems like plant-wide scheduling or supply chain optimization become computationally intractable due to their exponential growth in complexity. To address this, decomposition methods such as Benders decomposition prove invaluable by breaking the problem into more manageable master and subproblems, effectively separating strategic decisions from operational constraints.

Similarly, branch-and-bound algorithms help navigate the solution space more efficiently by prioritizing promising nodes and pruning redundant branches early in the search process. Another common pitfall arises with the use of big-M constraints, where overestimating the value of M can lead to weak relaxations and sluggish solver performance. The remedy lies in selecting the smallest physically meaningful M, such as deriving it from equipment capacity limits, or better yet, replacing big-M constraints entirely with specialized formulations such as SOS1/SOS2 constraints for piecewise linear approximations or indicator constraints supported by advanced solvers such as Gurobi and CPLEX.

Nonlinearities introduce additional complexity, particularly in MINLP problems involving reaction kinetics or thermodynamic relationships. Here, techniques such as piecewise linearization with SOS2 constraints or convex relaxations using McCormick envelopes can transform these nonlinearities into tractable MIP formulations. Symmetry in solutions, often encountered in problems with identical units such as parallel reactors, further complicates the solving process by creating redundant solution paths. Implementing symmetry-breaking constraints – such as enforcing an artificial ordering among variables – can significantly reduce the solver's workload. Finally, the challenge of finding initial feasible solutions can delay convergence, especially in tightly constrained systems. Warm-starting the solver with heuristic solutions or employing feasibility pumps to prioritize feasibility over optimality in early iterations can dramatically improve performance.

Ultimately, the key to successfully solving MIP problems in chemical engineering lies not just in selecting powerful solvers but also in crafting intelligent formulations that leverage process-specific insights. Tightening constraints, exploiting problem structure, and judiciously applying decomposition and linearization techniques often yield more substantial benefits than raw computational power. For practitioners, this means balancing mathematical rigor with a deep understanding of the underlying physical system to develop models that are both accurate and computationally efficient. Optimizing reactor networks, supply chains, or safety systems, these strategies collectively bridge the gap between theoretical complexity and practical solvability.

6.8 Superstructure Optimization

Superstructure optimization relies fundamentally on MIP to systematically evaluate discrete design choices and continuous operating variables in chemical process systems. The superstructure – a network encompassing all feasible unit operations, flow paths, and technologies – is formulated as an MINLP or MILP, where binary variables represent the presence or absence of equipment (e.g., reactors and heat exchangers) and integer variables quantify discrete decisions (e.g., number of stages in a distillation column). Continuous variables model operational parameters such as flow rates and temperatures. MIP enables rigorous handling of logical constraints (e.g., "if a compressor is installed, then its associated cooler must also be selected") through big-M formulations or disjunctive programming, while decomposition methods such as Benders decomposition and branch-and-bound algorithms tackle the combinatorial complexity inherent to large-scale superstructures.

The synergy between superstructure optimization and MIP is exemplified in problems such as heat exchanger network synthesis and process retrofitting, where binary variables activate/deactivate specific equipment or connections, and linear or nonlinear constraints enforce mass/energy balances and physical limits. For instance, selecting between alternative reactor designs can be encoded with binary variables, while their associated costs and performance are captured in the objective function. Advanced MIP techniques – such as McCormick envelopes for bilinear terms or SOS2 constraints for piecewise linear approximations – extend the approach to nonlinear problems. By leveraging MIP's ability to simultaneously optimize discrete and continuous decisions, superstructure methods provide a framework to navigate the trade-offs between capital expenditures, operating costs, and system performance, ultimately identifying globally optimal or near-optimal configurations for complex chemical processes.

6.9 Takeaway

In chemical engineering, we often face decision-making problems that contain discrete and continuous variables. Optimization models that contain integers (and continuous decisions) are called MIPs. In this chapter, we have provided a number of practical examples of MIPs, and we have discussed solution approaches to such problems: the branch-and-bound method and the cutting planes method. Both methods relax the integrality constraint and solve the resulting LP with a linear engine. In many cases, we are dealing with nonlinearities. We have discussed a number of techniques to linearize nonlinearities: the Glover linearization, the big-M method, and the lambda method (for piecewise linearization). We concluded the chapter with a discussion on logical inference, which is a powerful tool to translate heuristics into useful constraints.

6.10 Exercises

Exercise 1: ★ Basic: Equipment Selection Problem
A small biodiesel plant must choose between two reactor types:
– Continuous stirred-tank reactor: $120k capital, 8 tons/day capacity
– Plug flow reactor: $90k capital, 5 tons/day capacity

The plant needs to produce at least 20 tons/day to meet contracts. There's space for only one reactor type, but multiple units can be installed. The budget is $300k.

Tasks:
1. Formulate binary variables for reactor selection.
2. Use integer variables for number of units.
3. Minimize capital cost while meeting demand.

Learning objective: Basic binary/integer variable usage in equipment selection.

Exercise 2: ★★ Intermediate: Batch Scheduling with Changeovers
A pharmaceutical plant produces three antibiotics (A, B, C) in batches:

- Processing times: A (6 h), B (8 h), C (4 h)
- Changeover times: A→B (2 h), B→C (3 h), C→A (1 h)
- Demands: A (three batches), B (two batches), C (four batches)
- Equipment available 24/7 with one production line

Tasks:
1. Create time-indexed binary variables
2. Implement sequence-dependent changeovers
3. Minimize makespan

Learning objective: Discrete-time scheduling with sequence constraints.

Exercise 3: ★★ Advanced: Heat Exchanger Network Design
Retrofit a heat recovery system with the following:
- Two hot streams (H1: 300→200 °C, H2: 250→150 °C)
- Two cold streams (C1: 80→180 °C, C2: 100→200 °C)
- Three potential heat exchangers with different areas/costs
- ΔT_{min} = 20 °C required

Tasks:
1. Formulate binary variables for exchanger selection.
2. Implement temperature interval constraints.
3. Maximize heat recovery minus capital cost.

Learning objective: Network optimization with logical constraints.

Exercise 4: ★★★ Challenging: Refinery Crude Selection
A refinery can process four crude types with different properties:
- Crude A: $80/bbl, sulfur 1.2%, yield gasoline 40%
- Crude B: $70/bbl, sulfur 1.8%, yield gasoline 30%
- Crude C: $90/bbl, sulfur 0.8%, yield gasoline 50%
- Crude D: $75/bbl, sulfur 2.0%, yield gasoline 25%

Constraints:
- Total sulfur <1.5%
- Gasoline production >10,000 bbl/day
- Can process maximum two crude types simultaneously
- Minimum run length 3 days if selected

Tasks:
1. Formulate blending constraints.
2. Implement logic for minimum run lengths.
3. Maximize profit.

Learning objective: Complex blending with logical conditions.

Exercise 5: ★★ Intermediate: Safety System Configuration
Design a sensor network for a reactor system:
- Has five potential sensor locations.
- Each sensor covers specific combinations of three hazard zones.
- Cost ranges from $10k to $25k per sensor.
- Each zone must be covered by ≥2 sensors.
- If Sensor 1 is used, Sensor 3 cannot be used.

Tasks:
1. Create binary selection variables.
2. Formulate coverage constraints.
3. Implement mutual exclusion logic.
4. Minimize total cost.

Learning objective: Logical constraints in safety systems.

Exercise 6: ★★★ Advanced: Petrochemical Complex Planning
Optimize a five-process petrochemical complex over 12 months:
- Three production modes with different yields
- Two storage options with capacity limits
- Time-varying electricity prices
- Maintenance shutdown requirements
- Contractual demand fluctuations

Tasks:
1. Multiperiod MIP formulation
2. Inventory balance constraints
3. Mode selection logic
4. Minimize total cost

Learning objective: Large-scale multiperiod planning.

Exercise 7: ★ Basic: Utility System Optimization
Select between three boiler options:
- Boiler X: $500k capital, $50/MBtu

- Boiler Y: $700k capital, $40/MBtu
- Boiler Z: $300k capital, $60/MBtu

Requirements:
- Meet 100 MBtu/h base load
- Peak demand of 150 MBtu/h for ≤ 4 h/day
- Budget $1 million

Tasks:
1. Formulate cost function.
2. Implement load constraints.
3. Add binary selection variables.
4. Minimize 5-year cost.

Learning objective: Simple capital-operating cost trade-offs.

Group Project: Optimal Planning of a Multipurpose Batch Plant
1 Project Overview and Learning Objectives
In this 1-week team project, you will tackle a classic problem in chemical engineering operations: production planning with discrete decisions. You will formulate and solve an MILP model to schedule a batch plant that can produce multiple products using multiple equipment units.

By the end of this project, your team should be able to:
- Formulate a complex planning problem as an MILP, defining continuous variables, integer variables, linear constraints, and a linear objective function.
- Distinguish between continuous decisions (e.g., amounts and times) and discrete decisions (e.g., yes/no, assignments).
- Implement and solve the MILP model using specialized software (e.g., Python with PuLP/ORtools, MATLAB with intlinprog, GAMS, or Excel Solver).
- Interpret the integer solution and analyze its economic impact.
- Collaborate effectively to manage the tasks of model formulation, data handling, coding, and analysis.

2 Problem Statement
Your company operates a flexible batch plant with three reaction units (R1, R2, R3) and one blending/packaging unit (BPU). You must meet customer demand for four products (P1, P2, P3, P4) over the next 2 weeks (a 14-day planning horizon).

The catch: Not every reactor can make every product, and changing a reactor from one product to another requires a full day of cleaning, which costs money and loses production time.

The goal is to determine the optimal production schedule and product assignments to maximize total profit over the 2-week period.

Given Data:
- Demand: ($D_{p,d}$) kg of product p required on day d). A table will be provided (see Appendix A).
- Production Rates: Each reactor j has a specific production rate for each product p it can make ($R_{j,p}$ kg/day). A "–" means it cannot produce that product.

Unit	P1	P2	P3	P4	
R1	500	800	–	400	
R2	–	600	700	500	
R3	450	–	650	–	
BPU	1,200	1,200	1,200	1,200	*(The BPU can process all products.)*

Economic Data:
- Profit_p = Profit contribution for product p ($/kg):
  - P1: $$25$/kg, P2: $$30$/kg, P3: $$35$/kg, P4: $$20$/kg.
- ChangeoverCost = Cost to clean and change over a reactor to a different product = $$1,500$ per changeover.

Operational Rules:
- A reactor can only produce one product per day.
- The BPU can process the total output from all reactors each day (its capacity is not binding).
- Changeovers only happen *between* days. A reactor does not need a changeover to produce the same product on consecutive days.
- Initial state: At the start of Day 1, all reactors are clean and idle (no product assigned).

3 Mathematical Formulation (Your MILP Task)
Your team must define the decision variables, objective function, and constraints.
- Decision Variables:
  1. Continuous Variables:
     - $X_{j,p,d} \geq 0$ [kg]: Amount of product p produced in unit j on day d.
  2. Binary (Integer) Variables (0 or 1):
     - $Y_{j,p,d} \in \{0, 1\}$: 1 if unit j is *assigned* to product p on day d; 0 otherwise.
     - $Z_{j,p,q,d} \in \{0, 1\}$: 1 if unit j undergoes a changeover *from* product p^* *to* product q at the *start* of day d; 0 otherwise. *(Note: p and q can be the same, meaning no changeover).*
- Objective Function: Maximize Total Profit

  Maximize: Z = (Total Revenue from Sales) – (Total Changeover Costs)
  - Since demand must be met, revenue is fixed. Maximizing profit is equivalent to minimizing changeover costs. We can write:

    Maximize: $\Sigma (\Sigma \Sigma (\text{Profit\_p} * X_{j,p,d})) - \Sigma (\text{ChangeoverCost} * Z_{j,p,q,d})$.
  - *Sum over j, p, d for production; sum over j, p, q, d for changeovers.*
- Constraints:
  1. Demand Satisfaction: For each product p and each day d, the total produced must meet demand.
     $\Sigma_j X_{j,p,d} \geq D_{p,d}$ for all p, d
  2. Production Capacity: If a unit is assigned a product, it can produce up to its rate. If not, production must be zero. This is a logical constraint that links continuous and binary variables.
     $X_{j,p,d} \leq R_{j,p} * Y_{j,p,d}$ for all j, p, d
  3. Unit Assignment: Each reactor can be assigned to at most one product per day.
     $\Sigma_p Y_{j,p,d} <= 1$ for all j, d
  4. Changeover Logic: A changeover occurs in unit j on day d if the product assignment on day d-1 (p) is different from the assignment on day d (q).
     $Y_{j,p,d-1} + Y_{j,q,d} - 1 <= Z_{j,p,q,d}$ for all j, p, q, $d > 1$
     (This is a standard linearization for the condition IF (Y_{d-1} = 1 AND Y_d = 1) THEN Z = 1.)
  5. Non-negativity and Integrality: $X_{j,p,d} \geq 0$, $Y_{j,p,d}$ binary, $Z_{j,p,q,d}$ binary.

4 Deliverables and Schedule (1 Week)
- Day 1: Team formation and problem understanding. Discuss the key discrete decisions (assignments and changeovers). *Divide tasks: formulation, data setup, and coding research.*
- Day 2: Formulation Finalization. Whiteboard the entire MILP model. This is the most critical step.
- Days 3–4: Implementation. Code the model, input the demand data, and solve using an MILP solver. Debug the initial formulation.
- Day 5: Analysis and Reporting. Analyze the optimal schedule. Answer: How would the solution change if the changeover cost was only $$500$?

Final Submission (One PDF per Team):
1. Team Members and Contribution.
2. Complete MILP Formulation: Typed list of all variables, objective function, and constraints with a brief explanation of each.
3. Results:
    - The optimal total profit.
    - A Gantt chart or table showing the production schedule for reactors R1, R2, and R3 over the 14 days.
    - The total number of changeovers performed.
4. Discussion:
    - How does the model balance producing high-profit products with avoiding changeovers?
    - Was the solution found intuitive or surprising? Why?
    - What was the result of your sensitivity analysis on the changeover cost?
5. Appendix: Well-commented code and the input demand table.

5 Tips for Success
- Start Small: First, try solving the problem for a 2-unit, 2-product, 3-day horizon. This will help you debug your formulation quickly.
- Mind the Indices: MILP models have many indices (j, p, d). Be meticulous in your summation ranges and constraint definitions.
- Leverage Solver Logs: MILP solvers provide information on the solution gap. For a one-week project, it is acceptable to set a reasonable optimality gap (e.g., 1%) to get a good solution faster.
- Visualize the Answer: Creating a simple Gantt chart in Excel or Google Sheets is the best way to present and understand your final schedule.

Appendix A: Demand Table ($D_{p,d}$ in kg)

Day	P1	P2	P3	P4
1	200	0	400	100
2	0	500	300	200
3	400	200	0	150
...	...	...	...	...
14	300	100	250	0

(Note: You would provide a full 14-day table with realistic fluctuating demand).

Further Reading

Wolsey, L. A. (2020). *Integer programming* (2nd ed.). Springer.

Williams, H. P. (2013). *Model building in mathematical programming* (5th ed.). Wiley.

Nemhauser, G. L., & Wolsey, L. A. (1988). *Integer and combinatorial optimization*. Wiley.

Conforti, M., Cornuéjols, G., & Zambelli, G. (2014). *Integer programming*. Springer.

Jünger, M., Liebling, T. M., Naddef, D., Nemhauser, G. L., Pulleyblank, W. R., Reinelt, G., Rinaldi, G., & Wolsey, L. A. (Eds.). (2010). *50 years of integer programming 1958–2008: From the early years to the state-of-the-art*. Springer.

Google OR-Tools. (2023). *Mixed integer programming basics*. https://developers.google.com/optimization/mip/mip_intro

MIPLIB. (2017). *MIPLIB 2017: Mixed integer problem library*. http://miplib.zib.de/

Raman and Grossmann, (1991), Relation between MILP modelling and logical inference for chemical process synthesis, Computers & Chemical Engineering, 15(2), pp. 73–84.

7 Multi-objective Optimization

In optimization, the best answer is rarely a single point – it's a frontier of trade-offs, where every choice reveals what we truly value.

7.1 Introduction

Multi-objective optimization (MOO) is the science of making decisions when multiple, often competing, goals must be considered simultaneously. Unlike traditional single-objective problems, where the goal might be to minimize cost or maximize efficiency, MOO forces engineers to confront trade-offs inherent in real-world systems. For example, designing a chemical reactor might involve maximizing product yield while minimizing energy consumption, two objectives that frequently conflict. The challenge lies not in finding a single "best" solution, but in identifying a spectrum of optimal compromises that reflect the priorities of the decision-maker.

The importance of MOO in chemical engineering cannot be overstated. Process design, supply chain logistics, and sustainability initiatives all require balancing competing demands. A distillation column, for instance, might need to optimize for both operational cost and product purity, while staying within safety constraints. Similarly, a pharmaceutical manufacturer may need to weigh production speed against batch consistency. These problems share a common thread: improving one objective often comes at the expense of another, making MOO an essential tool for modern engineers.

One of the defining challenges of MOO is the absence of a unique solution. Instead, the outcome is typically a set of equally valid alternatives, each representing a different balance between objectives. This set, known as the *Pareto frontier*, graphically illustrates the trade-offs at play. For example, plotting energy use against product purity for a reactor design might reveal a curve where any improvement in purity requires greater energy expenditure. The role of the engineer is then to select the most appropriate solution from this frontier based on higher-level priorities, such as economic constraints or environmental regulations.

Real-world applications of MOO abound in chemical engineering. Petrochemical companies use it to optimize feedstock selection, balancing cost against processing complexity. Energy systems engineers rely on MOO to design renewable power grids that meet both cost and reliability targets. Even in emerging fields like carbon capture, MOO helps reconcile the competing demands of efficiency, scalability, and cost. A classic example is biodiesel production, where the choice between low-cost, high-impurity feedstocks and expensive but pure alternatives can significantly impact both profitability and environmental footprint.

Beyond technical considerations, MOO also raises broader questions about values and ethics in engineering. How much additional cost is justified to reduce emissions? Should a company prioritize short-term profits or long-term sustainability? These di-

https://doi.org/10.1515/9783111342283-007

lemmas underscore the fact that MOO is not just a mathematical exercise: it is a framework for making informed, responsible decisions in an increasingly complex world.

To illustrate, consider a simple case: optimizing a heat exchanger network for energy efficiency and capital cost. A more efficient design might require additional heat transfer area, driving up initial expenses. MOO provides the tools to quantify this trade-off, enabling engineers to present stakeholders with clear, data-driven choices. This interplay between technical rigor and strategic decision-making is what makes MOO both a powerful and indispensable discipline.

Case Example: Optimizing a Biofuel Production Process

Problem Statement:
A biorefinery aims to produce biodiesel from algae. Two key objectives conflict:
1. **Maximize annual production yield** (to boost revenue).
2. **Minimize water and energy consumption** (to reduce costs and environmental impact).

Trade-Off Analysis:
High-yield conditions require intensive water use for algae growth and significant energy for extraction. Low-resource conditions reduce operational costs but yield less product, hurting profitability.

MOO Approach:
Pareto Frontier Generated:
Plotting "production" versus "total energy cost" reveals a curve of optimal compromises (see Figure 7.1). If we were to optimize just for productivity, we find a productivity of 100 tons/day, with an energy requirement of 80 MWh/day. If we were to optimize just for energy, we find an energy requirement of ~ 5 MWh/day, with a productivity of ~20 tons/day.

Decision-Making Insight:
We can make trade-offs, for example point A, the knee point, where the productivity is 89.3 tons/day, and the energy use is 47.2 MWh/day (given a residence time of 6.7 h). If the focus is more toward energy savings, we could select point B (sustainability), where the energy use is 29.6 MWh/day, with a productivity of 71.2 tons/day, given a residence time of 4.6 h.

Outcome:
The MOO framework enables data-driven negotiation between production and sustainability teams, avoiding ad-hoc compromises.

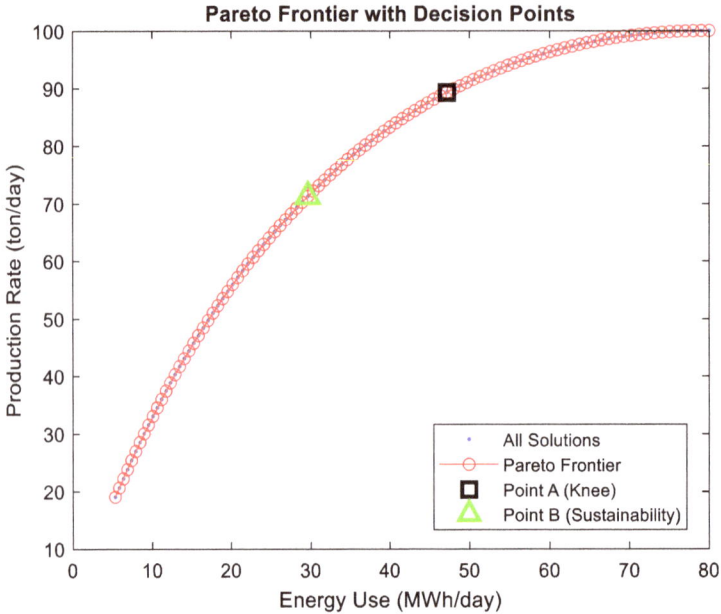

Figure 7.1: Pareto frontier with decision-points.

Discussion Questions

"What happens to Point B if the energy threshold is tightened?"

"How would a third objective (e.g., equipment wear) affect the Pareto frontier?"

7.2 Pareto Optimality and Dominance

7.2.1 The Concept of Pareto Optimality

In MOO, a solution is *Pareto optimal* if no objective can be improved without worsening at least one other objective. Imagine designing a chemical reactor where you aim to maximize yield while minimizing energy consumption. A design that achieves 90% yield with 50 MWh/day is Pareto optimal if no other design exists with both a higher yield and lower energy use. These solutions form the *Pareto frontier*, a curve (or surface) that visualizes the best possible trade-offs.

7.2.2 Dominance: Comparing Solutions

A solution *dominates* another if it is better in at least one objective and no worse in all others. For example:

- **Solution X**: 85% yield, 40 MWh/day.
- **Solution Y**: 90% yield, 45 MWh/day. Here, neither dominates the other because X uses less energy, but Y has a higher yield. However, both dominate a third solution, Z (80% yield, 50 MWh/day), which is inferior in both objectives. Dominance is the foundation for filtering out suboptimal solutions to construct the Pareto frontier.

7.2.3 Mathematical Formulation of MOO

A general MOO problem with k objectives and m constraints can be written as:

$$\min F(x) = |f_1(x),\ f_2(x),\ ...,f_k(x)|^T$$

Subject to $g_i(x) \leq 0,\ i = 1,\ ...,m$

$$h_j(x) = 0,\ j = 1,\ ...,p$$

$$x \in X \subseteq \mathbb{R}^n$$

where x are the decision variables (e.g., reactor temperature and flow rate), where $F(x)$ is a vector with the objectives (e.g., costs, emissions, and yield), and g_i, h_j are inequality and equality constraints, for example, safety limits, material balances, etc.

A solution x^* is called Pareto optimal is no other $x \in X$ satisfies (for minimization):

$$\forall i \in \{1,\ ...,k\}: f_i(x) \leq f_i(x^*)$$

In other words, these are the solutions on the Pareto frontier; any other solution above or below the frontier can be improved by moving on to the Pareto frontier.

7.2.4 Visualizing the Pareto Frontier

A 2D plot (e.g., energy use vs. production rate) effectively illustrates the Pareto frontier. The frontier's shape reveals the trade-off severity:

- **Convex Frontier**: Small sacrifices in one objective yield large gains in the other (e.g., slight energy increase drastically boosts yield).
- **Concave Frontier**: Significant compromises are needed for marginal improvements.

7.2.5 Limitations and Practical Nuances

- **Curse of dimensionality**: With > 3 objectives, visualizing the frontier becomes challenging, requiring dimensionality reduction techniques.
- **Subjectivity**: Pareto optimality identifies *possible* solutions but does not prescribe a single "best" choice, which depends on stakeholder priorities.

7.2.6 Pareto Optimality, Dominance, and Space Mapping

To fully grasp MOO, it is critical to distinguish between the decision space: the set of all possible values for the decision variables, and the objective space: the set of all possible objective function values. The transformation from decision to objective space is governed by:

$$F(x) = \begin{bmatrix} f_1(x) \\ \vdots \\ f_k(x) \end{bmatrix}, \quad x \in X$$

A single point in the decision space, maps to one point in the objective space. The Pareto frontier in the objective space corresponds to multiple optimal decisions in the decision space.

Let us consider an example:

7.2.6.1 The Production Planning Problem

We have two decision variables: x_1: units of product A to be produced (≥ 0) and x_2: units of product B to be produced (≥ 0)

We also pursue two objectives: $\max f_1(x_1, x_2) = 5x_1 + 3x_2$ (to maximize profit) and $\min f_2(x_1, x_2) = 2x_1 + 4x_2$ (to minimize labor hours)

In this program we have the following constraints: $3x_1 + 2x_2 \leq 18$ (A material constraint), $x_1 + 2x_2 \leq 12$ (Machine time), $x_1 \leq 5$ (Demand for A) and $x_1, x_2 \geq 0$ (nonnegativity).

We can plot the constraints in an x_1–x_2 plane. The corner points are (0,0), (0,6), (2,5), (5,1.5), and (5,0). The left figure on Figure 7.2 shows the decision space.

For each corner point we can now also calculate the objective function values:

If we plot f_1 versus f_2 we create the objective space. Figure 7.2, the right figure, shows the objective space.

x_1	x_2	f_1	f_2
0	0	0	0
0	6	18	24
2	5	19	24
5	1.5	29.5	16
5	0	25	10

The point (0,0) gives us the best solution for labor hours (0), with (0) profit. And the point (5,1.5) gives us the best solution for profit (29.5) with (16) labor hours.

Decision Space

Feasible Region
Material Constraint
Machine Constraint
Demand Limit

Feasible decisions → Objectives

x_1 (Units of Product A)

x_2 (Units of Product B)

Objective Space with Pareto Frontier

(16, 29.5)

(10, 25)

(0, 0)

Feasible Solutions
Pareto Frontier (Correct Order)

Labor Hours (1000s)

Profit ($1000s)

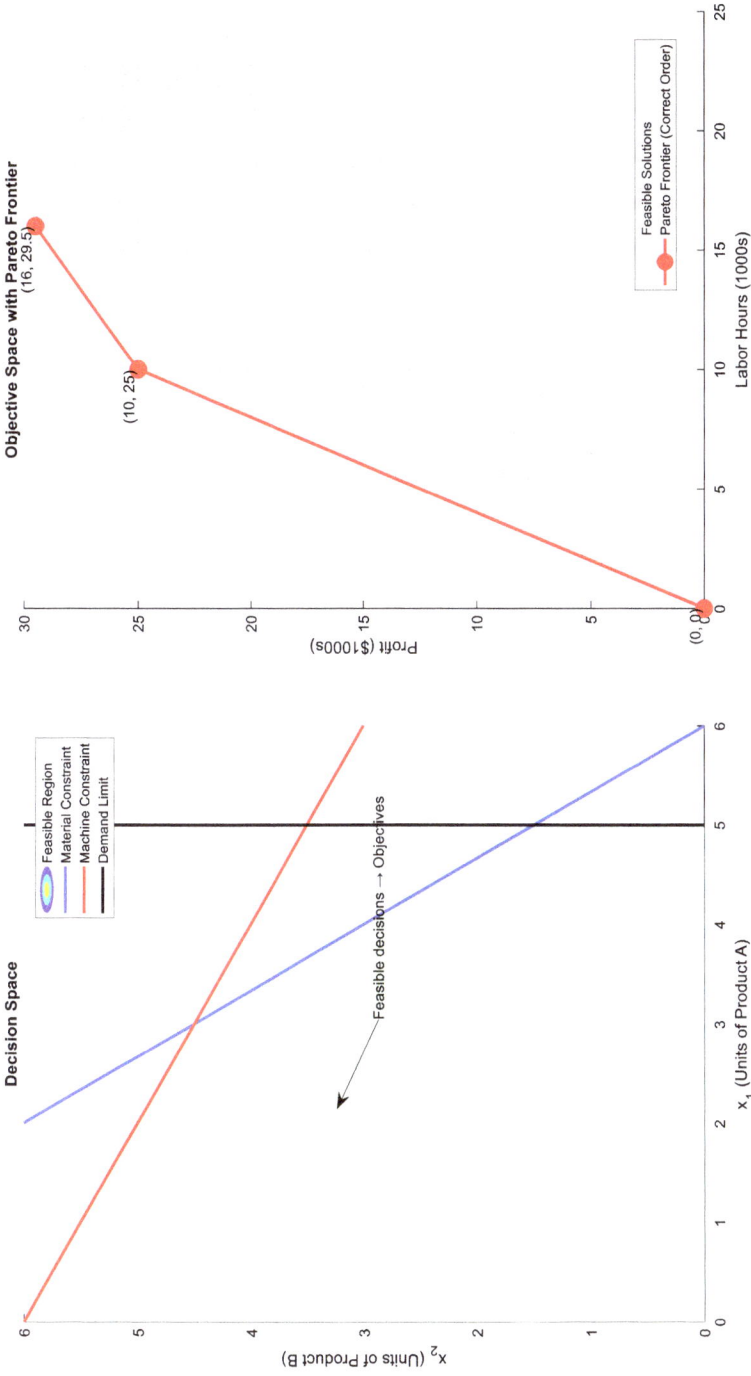

Figure 7.2: Decision space (left) and objective space (right).

All points in the shaded area are called dominated points; by moving up and/or to the left, we can find better solutions. Only the points on the red line segment are nondominated points; beyond this line, we cannot find any better solution other than improving one objective will deteriorate the other. We have to make trade-offs. This line segment is the Pareto frontier.

When we have more decision variables and more constraints, the mapping of the objective space becomes more and more difficult. Ultimately we are only interested in the Pareto frontier. There exist methods we can use to effectively generate the Pareto frontier. See in Figure 7.3 how PhD student Daan is making use of the pareto frontier to decide on multiple, conflicting objectives.

Figure 7.3: Grad student Daan and the Pareto Predicament.

7.3 Solution Methods for Multi-objective Optimization

As we have seen before, MOOproblems lack a single optimal solution but instead offer a set of Pareto-optimal solutions. In this section, we explore methods to identify and select solutions from the Pareto frontier, balancing competing objectives, like cost, efficiency, and sustainability.

7.3.1 Weighted Sum Method

In the *weighted sum method*, we combine objective into a single scalar function using weights:

$$\min \sum_{i}^{k} w_i f_i(x), \ w_i \geq 0, \sum w_i = 1$$

For our production planning problem, we could formulate an objective function as a weighted sum:

$$\min w_1(-f_1) + w_2 f_2$$

We use the minus sign since profit should be maximized.

We could try different values for the weights, for example $w_1 = 0.7$ and $w_2 = 0.3$, giving more weight to profit over labor hours. Or the other way around $w_1 = 0.3$ and $w_2 = 0.7$ (more weight to labor hours over profit).

Below you can find the complete MATLAB code that solves the LP for different weights with the `linprog` solver, and displays the outcomes:

```
% ====================================================================
% COMPLETE Multi-Objective Optimization Visualization
% Correct Pareto Frontier + Weighted Sum Solutions
% ====================================================================
clf; close all; clear all;

%% Problem Data
A = [3 2; 1 2; 1 0];      % Constraint matrix (Material, Machine, Demand)
b = [18; 12; 5];          % Constraint limits
lb = [0; 0];              % Lower bounds (x1, x2 >= 0)

%% Define Weight Combinations
weights = [0.9 0.1;       % 90% profit, 10% labor
```

```
            0.7 0.3;      % 70% profit, 30% labor
            0.5 0.5;      % 50% profit, 50% labor
            0.3 0.7];     % 30% profit, 70% labor


%% Compute All Feasible Points for Objective Space
[x1, x2] = meshgrid(linspace(0, 6, 100), linspace(0, 6, 100));
profit_all = 5*x1 + 3*x2;
labor_all = 2*x1 + 4*x2;
feasible = (3*x1 + 2*x2 <= 18) & (x1 + 2*x2 <= 12) & (x1 <= 5);


%% Compute CORRECT Pareto Frontier
corners = [0 0; 0 6; 2 5; 5 1.5; 5 0]; % Decision space corners
profit = 5*corners(:,1) + 3*corners(:,2);
labor = 2*corners(:,1) + 4*corners(:,2);


% Identify TRUE Pareto-optimal points (non-dominated)
pareto_mask = true(size(profit));
for i = 1:length(profit)
    if any((profit > profit(i)) & (labor <= labor(i))) || ...
        any((profit >= profit(i)) & (labor < labor(i)))
        pareto_mask(i) = false;
    end
end
pareto_points = [labor(pareto_mask), profit(pareto_mask)];
pareto_points = sortrows(pareto_points, 1); % Sort by labor


%% Compute Weighted Sum Solutions
solutions = zeros(size(weights,1), 5); % [w1, w2, labor, profit,
total_obj]
for i = 1:size(weights,1)
    f = -weights(i,1)*[5; 3] + weights(i,2)*[2; 4]; % Combined
objective
    [x_opt, ~] = linprog(f, A, b, [], [], lb);
    labor_opt = 2*x_opt(1) + 4*x_opt(2);
    profit_opt = 5*x_opt(1) + 3*x_opt(2);
    total_obj = weights(i,1)*profit_opt - weights(i,2)*labor_opt;
    solutions(i,:) = [weights(i,:), labor_opt, profit_opt, total_obj];
end


%% Figure 1: Objective Space with CORRECT Pareto Frontier
figure('Position', [100 100 700 600]);
```

```
hold on; grid on;

% Plot feasible region
scatter(labor_all(feasible), profit_all(feasible), 10, [0.8 0.9 1],
'filled', . . .
    'DisplayName', 'Feasible Region');

% Plot CORRECT Pareto frontier
plot(pareto_points(:,1), pareto_points(:,2), 'ko-', . . .
    'MarkerSize', 8, 'LineWidth', 1.5, 'MarkerFaceColor', 'k', . . .
    'DisplayName', 'Pareto Frontier');

% Label Pareto points
text(pareto_points(:,1), pareto_points(:,2), . . .
    {'(0,0)', '(10,25)', '(16,29.5)'}, . . .
    'VerticalAlignment', 'bottom', 'HorizontalAlignment', 'right');

% Plot weighted sum solutions
color_order = lines(size(weights,1));
for i = 1:size(weights,1)
    plot(solutions(i,3), solutions(i,4), 's', 'MarkerSize', 12, . . .
        'MarkerFaceColor', color_order(i,:), 'MarkerEdgeColor', 'k', . . .
        'DisplayName', sprintf('Weights %.1f:%.1f', weights(i,1),
weights(i,2)));
end

xlabel('Labor Hours (1000s)');
ylabel('Profit ($1000s)');
title('Objective Space with Correct Pareto Frontier');
legend('Location', 'southeast');
axis([0 25 0 35]);

%% Figure 2: Stacked Bar Chart (Correct Labels)
figure('Position', [800 100 700 500]);
hold on; grid on;

% Prepare data
profit_contrib = solutions(:,1) .* solutions(:,4);
labor_contrib = -solutions(:,2) .* solutions(:,3);
```

```
% Create stacked bars
b = bar([profit_contrib, labor_contrib], 'stacked');

% Color coding
b(1).FaceColor = [0.2 0.6 0.2]; % Green for profit
b(2).FaceColor = [0.8 0.2 0.2]; % Red for labor

% Create x-axis labels
labels = {
    'w_p=0.9, w_L=0.1'
    'w_p=0.7, w_L=0.3'
    'w_p=0.5, w_L=0.5'
    'w_p=0.3, w_L=0.7'
};

% Formatting
set(gca, 'XTick', 1:size(weights,1), . . .
         'XTickLabel', labels, . . .
         'TickLabelInterpreter', 'none');
ylabel('Objective Function Value');
title('Weighted Objective Contributions');
legend('Profit Contribution', 'Labor Contribution', 'Location',
'northoutside');

% Add total values
totals = solutions(:,5);
text(1:length(totals), totals, num2str(totals, '%.1f'), . . .
     'VerticalAlignment', 'bottom', 'HorizontalAlignment', 'center');
```

The advantages of the weighted sum method lie in the fact that it is simple to implement and it works very well for LP's. The main disadvantages are that it cannot capture concave Pareto frontiers, and the weight selection is subjective.

Not to mention possible imbalances in the terms of the objective function, suppose that profit is in the range of "millions" and hours are in the range of "thousands", a solver will in such a case, focus on the terms that have the largest contribution. To deal with such scaling problems, we might need to normalize each term in the objective function. There is also something fundamentally wrong with adding dollars to hours.

Objective Space with Correct Pareto Frontier

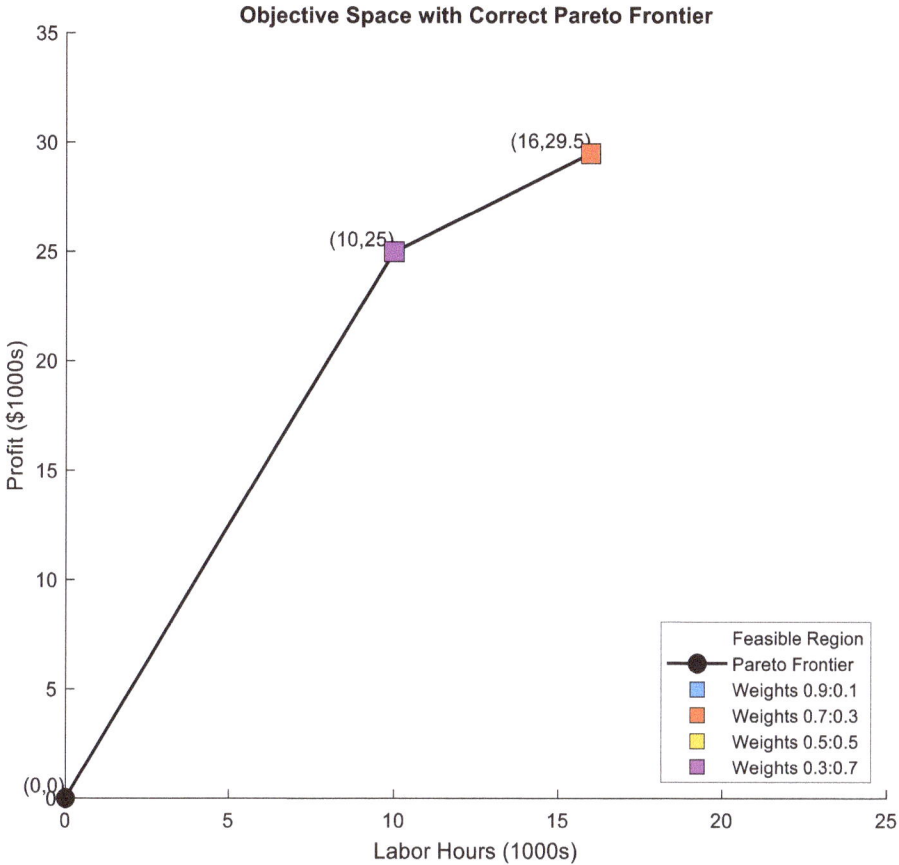

Figure 7.4: The objective space (shaded), the Pareto frontier (line) and the weighted sum solutions (squares).

Figure 7.4. shows the shaded objective space, the pareto frontier as a line and the weighted sum solutions (as squares).

This normalization can be achieved by first performing single objective optimizations to find optimal values for each of the individual objectives; the objective function could then be written as:

$$\min w_1 \frac{f_1}{f_1^*} + w_2 \frac{f_2}{f_2^*}$$

7.3.2 ε-Constraint Method

The *ε-constraint method* is a powerful technique for MOO that reformulates the problem by setting one primary objective to optimize and to convert the other objectives into constraints, with user-defined values (ε-values):

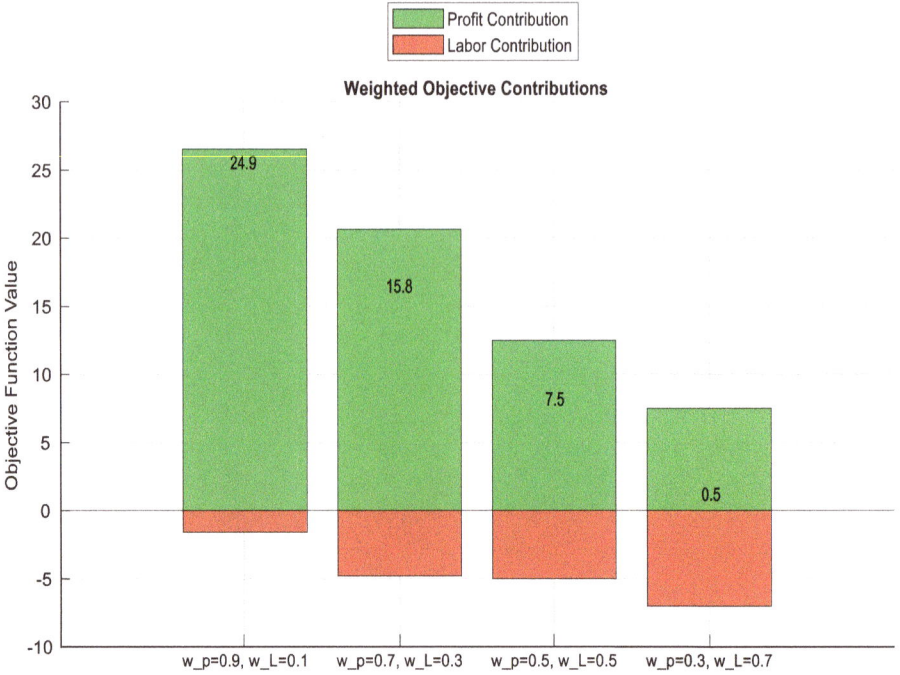

Figure 7.5: The objective values for the different weighted sum solutions, broken down into profit (green) and labor hours (red).

$$\min f_j(x)$$

Figure 7.5. shows the break down of the different weighted sum solutions in profit and labor hours.

Subject to:

$$f_i(x) \le \varepsilon_i \forall i \ne j$$

$$x \in X$$

where ε_i defines the maximum allowable value for the objective i. Let us try this out on our production planning problem:

$$\max 5x_1 + 3x_2$$

subject to:

$$2x_1 + 4x_2 \le \varepsilon$$

$$3x_1 + 2x_2 \le 18$$

$$x_1 + 2x_2 \le 12$$

$$x_1 \leq 5$$

$$x_1, x_2 \geq 0$$

Here we maximize the profit and use the labor hours as the ε-constraint. For different values of ε we get the following results:

ε	x_1	x_2	Profit
10.0	5	0	−25
13.5	5	0.875	−27.625
17.0	5	1.5	−29.5
20.5	5	1.5	−29.5
24.0	5	1.5	−29.5

With the MATLAB code below, these results can be generated:

```
% Define  ε  values for labor (sweep from min to max labor)
epsilon_values = linspace(10, 24, 5); % Example: 5 values between 10-24

% Initialize results
solutions = zeros(length(epsilon_values), 4); % [ ε , x1, x2, profit]

for i = 1:length(epsilon_values)
    % Add labor constraint: 2x1 + 4x2 <=  ε
    A_eps = [A; 2 4]; % Append labor constraint
    b_eps = [b; epsilon_values(i)];

    % Solve (maximize profit → minimize -profit)
    [x_opt, ~, exitflag] = linprog(-[5; 3], A_eps, b_eps, [], [], lb);

    if exitflag > 0% Solution found
        solutions(i,:) = [epsilon_values(i), x_opt', -[5 3]*x_opt];
    else
        solutions(i,:) = [epsilon_values(i), NaN, NaN, NaN]; %
Infeasible
    end
end

% Display results
disp(' ε -Constraint Results:');
disp(array2table(solutions, 'VariableNames', {' ε ', 'x1', 'x2',
'Profit'}));
```

```
% Plot ε-Constraint Results
figure;
plot(solutions(:,1), solutions(:,4), 'bo-', 'LineWidth', 1.5,
'MarkerSize', 8);
xlabel('Labor Limit (ε)');
ylabel('Maximum Profit');
title('Trade-off: Profit vs. Labor Constraint');
grid on;
```

Figure 7.6. shows the Pareto-frontier generated via the MATLAB code above.

From the expert: **Prof. Ana Povoa Barbosa**
Center for Management Studies
Instituto Superior Tecnico, University of Lisbon, Portugal

Designing and Planning Sustainable and Resilient Supply Chains

My personal interest in the study of sustainable and resilient supply chains was motivated by the fact that these systems are not merely logistical or economic mechanisms, but rather deeply interconnected networks that have a significant impact on society and the planet. Early in my research, I discovered that supply chains influence the extraction of resources, the consumption of products, and the management of waste. I recognized the urgent need to integrate sustainability into the core of supply chain decision-making and that sustainability cannot be fully realized without resilience, meaning without the capacity to anticipate, absorb, and adapt to disruptions.

One of the main achievements in this field has been the progressive integration of sustainability and resilience into supply chain design and the development of decision supporting tools to inform the complex decision process associated. Using multi-objective optimization and life cycle assessment techniques, the community has progressed from cost minimization to models that consider environmental and social impacts. Additionally, the formalization of the concept of resilience has led to the development of quantitative models that consider the ability of supply chains to recover from disruptions and adapt to new equilibria. It has been demonstrated that sustainable and resilient supply chains share fundamental principles: both require visibility, adaptability, and long-term thinking.

In particular, process supply chains are characterized by a close relationship between production and logistics. Such systems often have to deal with hazardous materials and must be able to manage continuous, batch and hybrid operations across multiple levels. They often operate under strong regulatory and environmental constraints and require high levels of capital investment. They must also adapt to fluctuating demand, variability in raw materials, and process uncertainty. Complexity is therefore the reality, and process system approaches have helped us to incorporate sustainability and resilience aspects into designs that allow for flexibility, redundancy, and rapid recovery in the face of uncertainty and risk without diminishing environmental and social concerns.

Yet, this continues to be an exciting field and significant challenges lie ahead. Despite progress, social sustainability remains underrepresented in modelling frameworks. Also, many resilience models still address resilience in a fragmented way, focusing on robustness or recovery speed while ignoring long-term adaptation, and there is often a lack of integration with sustainability dimensions. There is also a methodological gap: we need holistic, systems-level process models that account for the complexity of real supply chains, including forward and reverse flows, multiple tiers and interacting risks. Importantly, both resilience and sustainability require better metrics and decision support tools that are accessible to practitioners.

My advice to students entering this field is to learn to think in systems, not to forget process knowledge, and to act with purpose. Start with the big picture: where are the major impacts, biggest risks and unmet needs? Then use your process knowledge coupled with analytical skills to design robust, inclusive and adaptive solutions. Supply chains that are sustainable but not resilient will not endure. Likewise, process supply chains that are resilient but ignore sustainability will ultimately fail society. Your role is to bridge that gap with creativity, rigor, and responsibility.

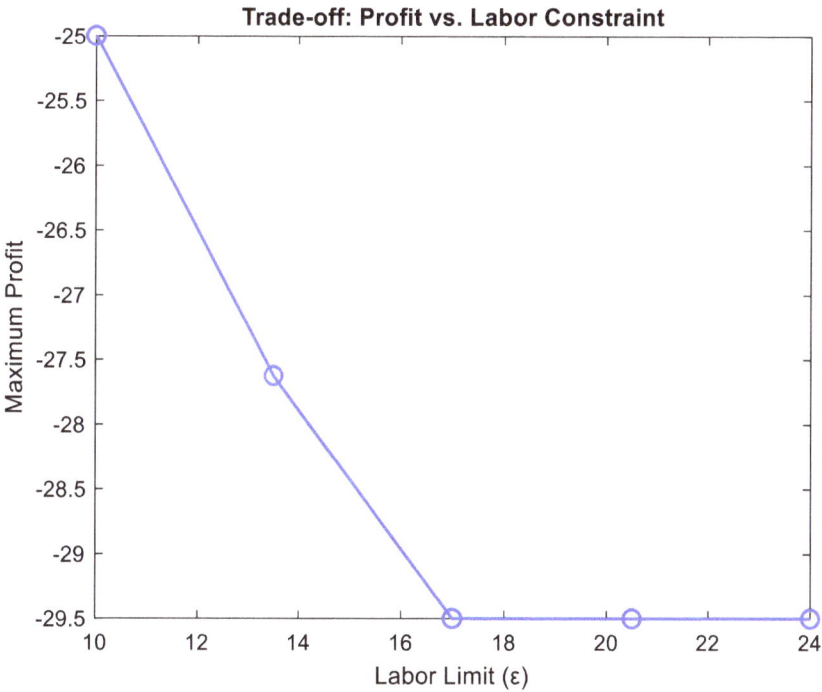

Figure 7.6: Trade-off of profit and labor hours.

The key properties of the ε-constraint method are:
1. Guaranteed Pareto optimality (every feasible solution is Pareto optimal).
2. Decreasing ε reduces the maximum achievable profit.
3. ε-Constraint works also with non-convex Pareto frontiers.

Important practical considerations:
1. Choosing ε values: You can start with bounds from single-objective optima.
2. Infeasibility handling: if ε is too small, the problem becomes infeasible.
3. Extensions: use adaptive ε sampling and/or applying to multiple objectives simultaneously.

7.3.3 Goal Programming

Goal programming is a MOO technique that aims to find solutions that minimize deviations from predefined target values for each objective. Unlike methods that seek to optimize objectives directly, goal programming focuses on achieving satisfactory performance levels, making it ideal for problems with conflicting goals or managerial targets.

The mathematical formulation of a goal program is for objectives $f_1, \ldots, f_k$ with targets $T_1, \ldots, T_k$:

$$\min \sum_i^k \left(d_i^+ + d_i^- \right)$$

subject to:

$$f_i(x) + d_i^- - d_i^+ = T_i \ \forall i$$

$$d_i^+, d_i^- \geq 0$$

$$x \in X$$

Here d_i^+ is a variable that allows a positive deviation (overachievement) and d_i^- is a variable that allows negative deviation (underachievement).

We can apply the goal programming concept to our production planning example. We target a profit of $T_1 = 25$ and a labor of $T_2 = 15$. The reformulated goal program is now:

$$\min \left(d_1^- + d_2^+ \right)$$

subject to:

$$5x_1 + 3x_2 + d_1^- - d_1^+ = 25$$

$$2x_1 + 4x_2 + d_2^- - d_2^+ = 15$$

$$3x_1 + 2x_2 \leq 18$$

$$x_1 + 2x_2 \leq 12$$

$$x_1 \leq 5$$

$$x_1, x_2, d_i^+, d_i^- \geq 0$$

It is noted that d_1^+ (profit overachievement) and d_2^- (labor savings) are not penalized in the objective. Below you find the worked-out MATLAB code that implements goal programming:

```
% ========================================================================
% Goal Programming Example
% Production Planning with Profit/Labor Targets
% ========================================================================

% Problem data (original constraints)
A_original = [3 2;     % Material: 3x1 + 2x2 <= 18
              1 2;     % Machine: x1 + 2x2 <= 12
              1 0];    % Demand: x1 <= 5
b_original = [18; 12; 5];
lb = [0; 0];           % x1, x2 >= 0

% Goal targets
profit_target = 25;    % Desired profit ($1000s)
labor_target = 15;     % Desired labor (1000 h)

% --- Reformulate as Goal Program ---
% Variables: [x1; x2; d1-; d1+; d2-; d2+]
% d1- = profit shortfall, d1+ = profit excess
% d2- = labor savings, d2+ = labor excess

% Objective: Minimize deviations (prioritize profit)
f = [0; 0;     % x1, x2 (no direct cost)
     1; 0;     % Minimize d1- (profit shortfall)
     0; 1];    % Minimize d2+ (labor excess)

% Goal constraints (equalities)
Aeq = [5 3 -1 1 0 0;   % 5x1 + 3x2 + d1- - d1+ = profit_target
       2 4 0 0 -1 1];  % 2x1 + 4x2 + d2- - d2+ = labor_target
beq = [profit_target; labor_target];

% Combine original + non-negativity constraints
A = [A_original zeros(3,4);   % Original constraints (no
deviation vars)
```

```
       0 0 1 0 0 0;           % d1- >= 0
       0 0 0 1 0 0;           % d1+ >= 0
       0 0 0 0 1 0;           % d2- >= 0
       0 0 0 0 0 1];          % d2+ >= 0
b = [b_original; zeros(4,1)];


% Solve
[x_opt, ~, exitflag] = linprog(f, A, b, Aeq, beq, [lb; zeros(4,1)]);


% --- Results ---
if exitflag > 0
    x1 = x_opt(1); x2 = x_opt(2);
    d1_neg = x_opt(3); d1_pos = x_opt(4);
    d2_neg = x_opt(5); d2_pos = x_opt(6);

    fprintf('Optimal Solution:\n');
    fprintf('x1 = %.2f, x2 = %.2f\n', x1, x2);
    fprintf('Profit = $%.2fK (Target: $%.2fK)\n', 5*x1 + 3*x2,
profit_target);
    fprintf('Labor = %.2fK hours (Target: %.2fK)\n', 2*x1 + 4*x2,
labor_target);
    fprintf('Deviations:\n');
    fprintf(' Profit shortfall (d1-) = %.2f\n', d1_neg);
    fprintf(' Labor excess (d2+) = %.2f\n', d2_pos);
else
    error('No feasible solution found! Relax targets.');
end
```

We would find for $x_1 = 3.93$ and $x_2 = 1.79$ leading to a profit of \$25.00 K and a labor of 15.00 K hours. In this case the deviations to the targets are zero.

7.3.4 Evolutionary Algorithms

Evolutionary algorithms (EAs) are population-based metaheuristics inspired by biological evolution. EA's can be very effective for solving complex MOO problems. A widely used EA is the non-dominated Sorting Genetic Algorithm II (NSGA-II). Figure 7.7 shows the algorithm flowchart.

Evolutionary algorithms like NSGA-II employ non-dominated sorting to rank solutions into Pareto fronts and crowding distance to maintain diversity, using genetic operators (selection, crossover, and mutation) to explore the search space. Unlike classical methods, EAs excel at handling non-convex Pareto fronts, discrete variables, and parallelization, though they require careful tuning of population size, crossover/mutation rates, and constraint-handling techniques (e.g., penalty functions). Their population-based approach avoids local optima but comes with higher computational costs, making them ideal for complex problems where gradient-based methods fail.

In practice, NSGA-II parameters must balance exploration and exploitation: typical settings include a population size of 50–200, crossover rates of 0.7–0.9, and mutation rates inversely proportional to variable count. Constraints are often managed via penalties (e.g., quadratic penalties for violations). However, EAs lack convergence guarantees and may stall on expensive simulations. For example, in production planning, NSGA-II approximates the trade-off between profit and labor but could miss exact Pareto points without sufficient generations or diversity maintenance.

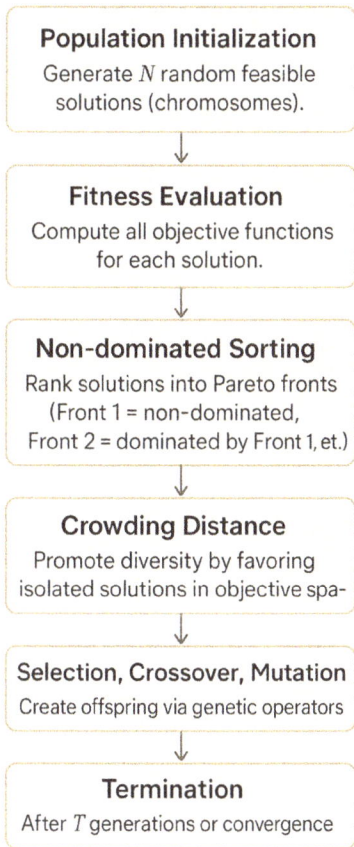

Population Initialization
Generate N random feasible solutions (chromosomes).

↓

Fitness Evaluation
Compute all objective functions for each solution.

↓

Non-dominated Sorting
Rank solutions into Pareto fronts (Front 1 = non-dominated, Front 2 = dominated by Front 1, et.)

↓

Crowding Distance
Promote diversity by favoring isolated solutions in objective spa–

↓

Selection, Crossover, Mutation
Create offspring via genetic operators

↓

Termination
After T generations or convergence

Figure 7.7: Algorithm overview for the NSGA-II.

To address limitations, hybrid EAs combine global search with local optimization (e.g., pattern search), while NSGA-III extends the framework to many-objective problems (>3 objectives). Surrogate models (e.g., kriging) accelerate optimization by approximating costly objectives. Industrial applications increasingly leverage GPU acceleration and distributed computing to scale EAs for high-dimensional problems, though algorithmic advances remain critical for robust performance. Future work may integrate machine learning for adaptive parameter tuning or real-time constraint handling.

MATLAB's gamultiobj is a built-in function that implements the NSGA-II evolutionary algorithm for solving MOO problems. It evolves a population of solutions over generations to approximate the Pareto frontier.

Below you can find the MATLAB implementation of the NSGA-II algorithm for solving the production planning problem:

```
% =========================================================================
% NSGA-II for Production Planning
% =========================================================================

% Problem definition
fitnessfcn = @(x) [-5*x(1) - 3*x(2); % Profit (maximized as -f1)
                   2*x(1) + 4*x(2)]; % Labor

% Constraints (A*x <= b)
A = [3 2; 1 2; 1 0];
b = [18; 12; 5];
lb = [0; 0]; % x1, x2 >= 0

% NSGA-II options
options = optimoptions('gamultiobj', ...
    'PopulationSize', 100, ...
    'ParetoFraction', 0.7, ...
    'MaxGenerations', 50, ...
    'CrossoverFraction', 0.8, ...
    'PlotFcn', @gaplotpareto);

% Run optimization
[x_opt, f_opt] = gamultiobj(fitnessfcn, 2, A, b, [], [], lb, [],
options);

% Post-processing
pareto_profit = -f_opt(:,1); % Convert back to profit
pareto_labor = f_opt(:,2);
```

```
figure;
scatter(pareto_labor, pareto_profit, 'filled');
xlabel('Labor (K hours)'); ylabel('Profit ($K)');
title('NSGA-II Pareto Front');
```

Figure 7.8 shows the Pareto frontier with gamultiobj.

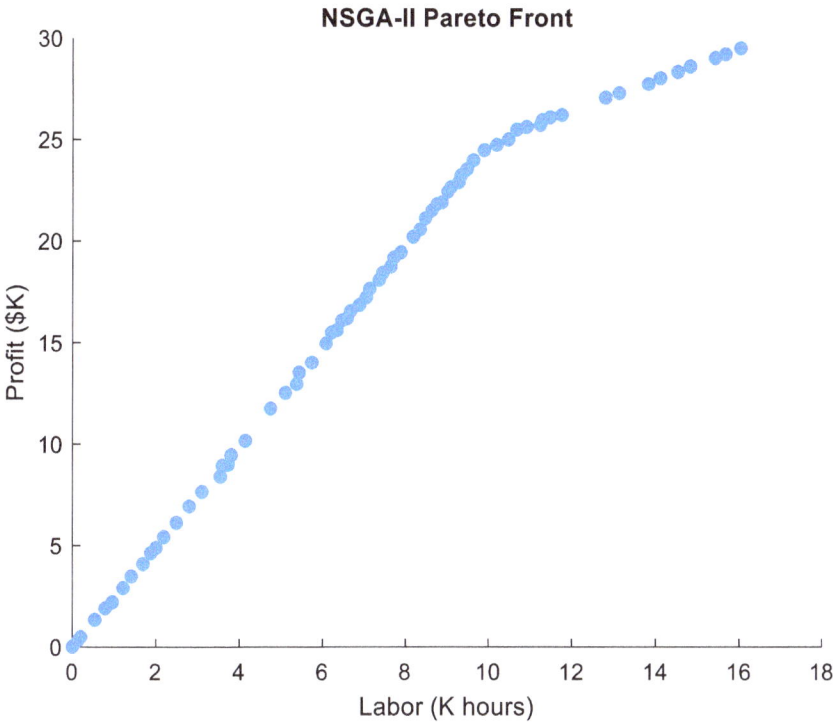

Figure 7.8: NSGA-II Pareto frontier for the production planning problem.

7.4 Decision Making in MOO

Selecting a final solution from the Pareto frontier requires balancing competing objectives with stakeholder priorities. Common approaches include a priori methods (e.g., weighted sums with fixed preferences) and a posteriori methods (e.g., analyzing the Pareto frontier first and then choosing). Visualization tools like parallel coordinate plots or interactive trade-off maps help decision-makers explore compromises.

For chemical engineers, decision-making often incorporates safety, sustainability, and economic constraints beyond technical performance. Methods like the technique for order preference by similarity to ideal solution or utility functions quantify preferences, while robust optimization accounts for uncertainty in parameters like feed-stock prices or demand fluctuations.

7.5 Applications in Chemical Engineering

MOO is widely used in process design, such as optimizing reactor yield versus energy consumption or distillation column efficiency versus capital cost. Case studies include biodiesel production (profit versus carbon footprint) and pharmaceutical batch processes (purity versus production time).

Emerging applications integrate MOO with machine learning for real-time optimization of smart plants. Examples include catalyst design (activity versus stability) and supply chain logistics (cost versus resilience), where Pareto analysis reveals non-intuitive trade-offs validated through digital twins.

7.6 Practical Implementation

Software tools like MATLAB's gamultiobj, Python's PyMOO, or Aspen Plus's optimization modules streamline MOO implementation. Key steps include problem formulation (objective/constraint definition), solver selection (evolutionary algorithms for non-convex problems), and post-processing (Pareto visualization).

Challenges include computational cost for large-scale systems and interpreting high-dimensional Pareto fronts. Best practices involve sensitivity analysis (e.g., Monte Carlo sampling) and embedding domain knowledge (e.g., prioritizing safety constraints) to reduce the solution space.

7.7 Challenges and Future Directions

Current limitations include handling many-objective problems (>4 objectives) and uncertainty quantification in dynamic environments. Scalability remains an issue for high-fidelity models like CFD-coupled reactor simulations.

Future trends focus on AI-augmented MOO, such as reinforcement learning for adaptive weight selection, and quantum computing to accelerate evolutionary algorithms. Ethical considerations (e.g., bias in sustainability metrics) and explainable AI for Pareto decisions are also critical research areas.

7.8 Takeaways

This chapter introduced MOO methods for chemical engineering, emphasizing trade-offs between conflicting objectives like cost and sustainability. We defined Pareto optimality and covered several solution techniques (weighted sum, ε-constraint, and goal programming) and decision-making frameworks. We highlighted applications from reactor design to supply chains, and outlined software tools and implementation challenges. Evolutionary algorithms (e.g., NSGA-II) were presented as robust solvers for non-convex problems. Key hurdles include computational expense and many-objective optimization, addressed through emerging technologies like AI and quantum computing. The chapter underscored the need for domain knowledge in interpreting Pareto frontiers and the growing role of digital twins. Practical examples reinforced how MOO balances technical and economic constraints. Future directions prioritize ethical AI and explainable results. Overall, MOO is a transformative tool for sustainable process engineering.

Further Reading

Cohon, J. L. (1978). *Multiobjective programming and planning*. Academic Press.

Deb, K., Pratap, A., Agarwal, S., & Meyarivan, T. (2002). A fast and elitist multiobjective genetic algorithm: NSGA-II. *IEEE Transactions on Evolutionary Computation*, 6(2), 182–197.

Marler, R. T., & Arora, J. S. (2004). Survey of multi-objective optimization methods for engineering. *Structural and Multidisciplinary Optimization*, 26(6), 369–395.

Rangaiah, G. P., & Bonilla-Petriciolet, A. (Eds.). (2013). *Multi-objective optimization in chemical engineering*. Wiley.

Zhang, Q., & Li, H. (2007). MOEA/D: A multiobjective evolutionary algorithm based on decomposition. *IEEE Transactions on Evolutionary Computation*, 11(6), 712–731.

7.9 Exercises

Exercise 1: Pareto Frontier Identification
Difficulty: ★
Problem:
Three reactor designs have the following performance:
– Design A: Profit = \$22K, Energy Use = 50 MWh
– Design B: Profit = \$25K, Energy Use = 60 MWh
– Design C: Profit = \$20K, Energy Use = 40 MWh

Tasks:
1. Identify which designs are Pareto-optimal.
2. Plot the solutions in objective space and shade the dominated region.

Exercise 2: Weighted Sum Method
Difficulty: ★★
Problem:
Optimize the production planning problem (from Section 7.3) with:
– Objective 1: Maximize profit $5x1+3x25x1+3x2$.
– Objective 2: Minimize labor $2x1+4x22x1+4x2$.
– Constraints: $3x1+2x2\leq183x1+2x2\leq18$, $x1+2x2\leq12x1+2x2\leq12$, $x1\leq5x1\leq5$.

Tasks:
1. Solve using weights $w1=0.7w1=0.7$ (profit), $w2=0.3w2=0.3$ (labor).
2. Report the solution and compute both objectives.

Exercise 3: ε-Constraint Trade-Off Analysis
Difficulty: ★★
Problem:
Using the same production planning problem:
1. Maximize profit while limiting labor $\leq\epsilon\leq\epsilon$ for $\epsilon=10, 12, 14, 16\epsilon=10, 12, 14, 16$.
2. Plot the resulting profit versus $\epsilon\epsilon$ and identify the "knee" point.

Exercise 4: Goal Programming Formulation
Difficulty: ★★★
Problem:
Reformulate the production problem with:
– **Goal 1**: Profit $\geq$\$24K.
– **Goal 2**: Labor $\leq$14K h.
– **Priority**: Avoid profit shortfall twice as much as labor excess.

Tasks:
1. Write the complete goal programming formulation.
2. Solve using MATLAB/Python and report deviations.

Exercise 5: NSGA-II Implementation
Difficulty: ★★★★
Problem:
Use MATLAB's gamultiobj to solve the production problem with:
– Objectives: Maximize profit, minimize labor.
– Constraints: Original problem constraints.

Tasks:
1. Run NSGA-II with a population of 100 for 50 generations.
2. Plot the Pareto frontier and compare to the exact frontier from Section 7.2.

Exercise 6: Real-World Ethical Trade-Offs
Difficulty: ★★
Problem:
A chemical plant must balance:
– **Profit**: $30K (target).
– **CO_2 Emissions**: ≤100 kg (target).

Tasks:
1. Propose a MOO method to address this.
2. Discuss how weights/constraints could reflect ethical priorities.

Multi-objective Optimization (MOO) Team Project
Title: *Design Your Own Optimization Problem*
Duration: 1 week (4–5 h of teamwork)
Team Size: 4 students
Deliverables:
1. A well-defined LP/MOOP problem statement (PDF).
2. MATLAB/Python code implementing the solution.
3. Pareto frontier visualization and trade-off analysis.
4. 10-minute presentation justifying decisions.

Project Phases
Phase 1: Problem Definition (Day 1)
Task:
Use the provided **parameter table** (below) to design your own LP/MOOP with:
– **2–3 objectives** (e.g., maximize profit and minimize waste).
– **3–4 constraints** (e.g., budget and resource limits).
– **2–3 decision variables** (e.g., production rate and staffing).

Parameter Table (Generic Industrial Process):

Parameter	Symbol	Value range	Units
Raw material cost	$c1$	10–50	$/kg
Energy cost	$c2$	0.5–2.5	$/kWh
Production rate	$x1$	100–1,000	kg/day
Labor hours	$x2$	8–24	h/day
Yield efficiency	η	70–95%	–
CO_2 emissions	e	0.1–5.0	kg/kg output
Max budget	–	$50,000	$
Max energy usage	–	2,000	kWh/day

Example Objectives:
- Maximize profit: Revenue–($c1x1$+$c2$Energy).
- Minimize CO_2: $e{\times}x1$.
- Maximize yield: $\eta{\times}x1$.

Example Constraints:
- Total cost≤$50,000.
- Labor hours≤24.

Phase 2: Model Implementation (Days 2–3)
Tasks:
1. Formulate your LP/MOOP mathematically (define objectives/constraints).
2. Solve using:
    - **Weighted sum method** (vary weights for objectives).
    - **ε-Constraint method** (fix one objective, optimize the other).
3. Generate Pareto frontiers.

Tools:
- MATLAB (linprog, gamultiobj).
- Python (scipy.optimize, PyMOO).

Phase 3: Sensitivity Analysis (Day 4)
Tasks:
1. Test how solutions change with ±20% parameter variations (e.g., raw material cost).
2. Identify "brittle" solutions (small changes cause large performance drops).

Phase 4: Reporting (Days 5–7)
Report Sections:
1. **Problem Statement**: Your custom LP/MOOP.
2. **Methods**: Solvers and parameters used.
3. **Results**: Pareto plots and sensitivity tables.
4. **Discussion**: Trade-offs and recommendations.

Presentation:
– 5 slides max.
– Focus on **how you designed the problem** and key insights.

8 Decision-Making Under Uncertainty

Uncertainty is the only certainty in chemical processes. The rest is optimization.

8.1 Introduction

Optimization in chemical engineering often assumes a world of perfect knowledge, fixed parameters, predictable demands, and deterministic outcomes. Yet, reality is far messier. Feedstock prices fluctuate, reactor yields vary, equipment fails, and market demands shift overnight. Traditional optimization methods, while powerful, can deliver fragile solutions when confronted with such uncertainty.

This chapter tackles a critical question: *How can we optimize processes when the inputs themselves are uncertain?* Whether designing a plant, planning supply chains, or scheduling production, ignoring uncertainty risks costly overdesign, operational disruptions, or missed opportunities. Here, we move beyond deterministic models to tools that embrace variability, stochastic programming, robust optimization, and chance constraints, equipping you to make decisions that are not just optimal *on paper*, but *in practice*.

We begin by exploring real-world sources of uncertainty and then formalize mathematical frameworks to quantify and manage risk. Case studies and software tools will bridge theory to industrial applications, demonstrating how embracing uncertainty can turn volatility into a competitive edge.

8.2 Sources of Uncertainty in Chemical Processes

Process optimization under uncertainty begins with understanding *where* and *why* variability arises in chemical engineering systems. Unlike textbook problems, real-world processes operate amid ever-changing conditions, some quantifiable, others unpredictable. This section categorizes the key sources of uncertainty, emphasizing their impact on optimization outcomes.

Process Variability is the inherent fluctuations in physical, chemical, or operational parameters. Process variability includes the variability in raw material purity (e.g., crude oil grades and biomass moisture content) that affects reaction yields and separation efficiency, time-dependent activity loss alters reactor performance, requiring dynamic optimization, and/or sensor errors that propagate through control systems, distorting real-time optimization. An optimization challenge here is that deterministic

https://doi.org/10.1515/9783111342283-008

models assume fixed parameters, but variability can lead to suboptimal or infeasible solutions.

Market uncertainty concerns the external economic factors that influence process economics. You can think of fluctuations in energy costs (e.g., natural gas) or product demand (e.g., seasonal pharmaceuticals) impacting profitability, and/or new sustainability mandates (e.g., carbon taxes) may abruptly change cost structures. In deterministic optimization, the static cost functions fail to account for future market scenarios, risking over-/underinvestment.

Operational risk includes unplanned disruptions in process execution. For example, the pump/compressor downtime reduces throughput, requiring redundant capacity, or think of delays in raw material delivery (e.g., geopolitical events) that force production rescheduling. Traditional scheduling models assume continuous operation, ignoring downtime penalties.

Environmental and external factors from uncontrollable external influences, think of renewable energy availability (solar/wind) that fluctuates, affecting energy-intensive processes or policy changes that target emission limits, leading to retrofitting costs into existing designs. Long-term projects (e.g., plant design) must hedge against "unknown unknowns."

And lastly, **model uncertainty** stems from gaps between theoretical models and real-system behavior. For example, approximate reaction rate expressions lead to inaccurate reactor sizing, or as a second example, nonideal mixture behavior in distillation column performance. Optimal solutions based on imperfect models may perform poorly in practice.

To illustrate, consider a **case vignette:**

A biodiesel plant optimizes production assuming fixed feedstock (soybean oil) costs. When a drought spikes prices by 40%, the "optimal" design becomes uneconomical. Had the model incorporated price distributions, it might have diversified feedstocks (e.g., waste cooking oil) for resilience.

8.3 Mathematical Frameworks for Uncertainty

Optimization under uncertainty requires moving beyond deterministic models to a setting that explicitly accounts for variability, risk, and incomplete knowledge. This section introduces three concepts tailored to different types of uncertainty and decision-making contexts: stochastic programming, robust optimization, and chance-constrained programming.

Let us take a look at a simple refinery production problem. We want to process two types of crude into two types of products (say gasoline and fuel oil). We know the costs of the raw materials, the demand for products, the production capacity, and the productivity. We are interested in finding the quantities of raw materials that should be processed in such a way that the costs are minimal.

With our knowledge of linear programming, we set up the following LP:

$$\min Z = 2x_1 + 3x_2$$

subject to:

$$x_1 + x_2 \leq 100$$

$$2x_1 + 6x_2 \geq 180$$

$$3x_1 + 3x_2 \geq 162$$

$$x_1, x_2 \geq 0$$

The first constraint denotes the capacity, and the following two constraints represent the demand for the two different products and their productivity. x_1 and x_2 are the quantities of raw materials processed?

We can use the Simplex method or the graphical method to find the optimum. In Figure 8.1, the decision space and the optimum are drawn.

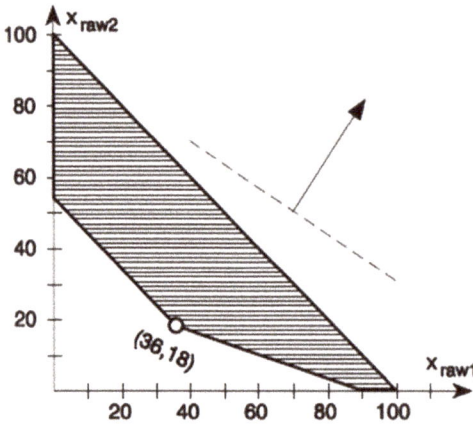

Figure 8.1: The shaded area is the feasible region. At point (36,18), the optimum is found.

From the expert: **Prof. Efstratios Pistikopoulos**
Texas A&M University, USA

Advances in Multiparametric Programming

Multiparametric Programming: Multiparametric programming is an advanced mathematical optimization technique for analyzing the effect of varying parameters and/or uncertainties in a systematic manner. Essentially it identifies the partitions of the parameter space into areas, known as critical regions. Each of these regions are defined by explicit functions that describe the optimization problem in terms of its bounded parameters.

Explicit Multiparametric Model Predictive Control (MP-MPC): One of the most important applications of multiparametric programming is its use in model-base control, specifically, model-predictive control (MPC). By setting the system-state variables/process data as bounded parameters, explicit functions of the optimal control inputs can be determined using multiparametric programming. Implementing these control actions on a standard micro-chip enables the application of the "MPC-on-a-chip" technology.

PAROC Framework: As most real-world systems are a culmination of optimization and modelling tasks such as parameter estimation, dynamic optimization, design, scheduling and control, the underlying models that represent such systems become challenging to handle within the scope of MPC. The PARametric Optimization and Control (PAROC) framework was developed in light of this challenge. The framework in brief initially calls for the development of "high-fidelity" model, based on which an approximate model is developed using system identification and model reduction techniques. Multiparametric programming is then applied to this model to obtain its explicit control laws, which are then finally validated against the original high-fidelity model.

Key Developments: Theoretical and practical advances in multiparametric programming for energy and process systems engineering, biological and biomedical systems engineering, sustainable smart manufacturing and process intensification.

References

[1]   E. Pistikopoulos, Multi-parametric optimization and control, Wiley (2020)
[2]   S. Avraamidou and E. Pistikopoulos, Multi-level mixed-integer optimization – parametric programming approach, de Gruyter (2022).
[3]   B. Burnak, N. Diangelakis, E. Pistikopoulos, Integrated process design and operational optimization via multi-parametric programming, Springer (2020)
[4]   E. Pistikopoulos and Y. Tian, Synthesis and operability strategies for computer aided molecular process intensification, Elsevier (2022)

We have found the solution to the problem, provided that productivity, costs, demand, and capacity are fixed data, known to us prior to our optimization efforts. In this case, we speak of a deterministic problem.

However, at least some of the data (productivity and demand, for instance) may vary within certain limits, and we have to make our decisions on the production plan before knowing the exact values of those data. In other words, our original LP does no longer reflect the real problem.

We could adapt our LP:

$$\min Z = 2x_1 + 3x_2$$

subject to:

$$x_1 + x_2 \leq 100$$

$$(2 + \eta_1)x_1 + 6x_2 \geq 180 + \zeta_1$$

$$3x_1 + (3 + \eta_2)x_2 \geq 162 + \zeta_2$$

$$x_1, x_2 \geq 0$$

where the ζ and η are distributions, i.e., $\zeta_1 \in [-30.91, \ 30.91]$, $\zeta_2 \in [-23.18, 23.18]$, $\eta_1 \in [-0.8, \ 0.8]$ and $\eta_2 \in [0, \ 1.84]$. Please note that the distributions are not necessarily symmetric.

The consequence of variations in the distributions is that the feasible space will look different. In Figure 8.2 we can see how different demands *translate* the decision space. Figure 8.3 shows how different productivities *rotate* the decision space.

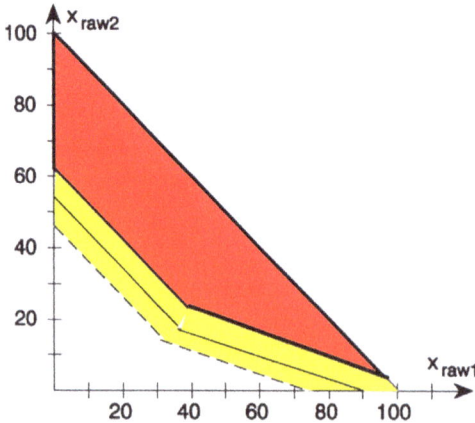

Figure 8.2: Feasible space changes with different demands.

The optimal solution lies elsewhere, depending on the actual realizations of the uncertainty.

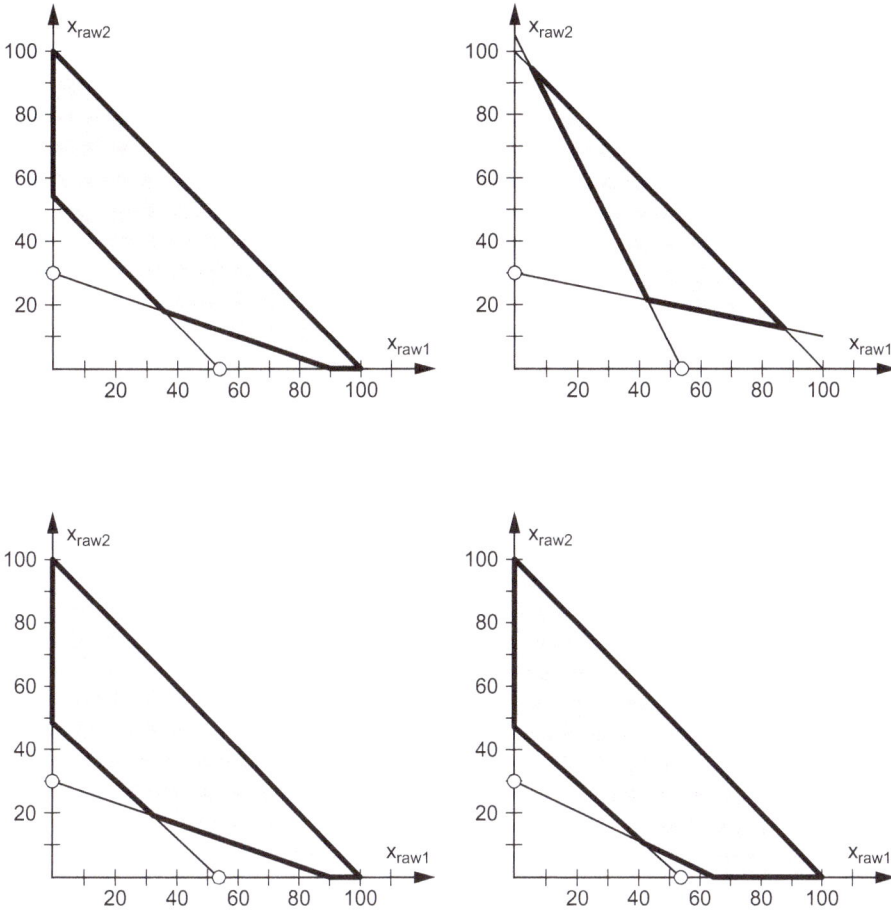

Figure 8.3: Feasible space changes with different productivities.

8.3.1 Stochastic Programming

We can use stochastic programming when the uncertainty can be described by known probability distributions (as in the previous example). One way is via *scenario-based modeling*, where we discretize continuous uncertainty into finite scenarios (e.g., low/medium/high demand) to each of these scenarios we assign a probability.

Often the problem is broken up in to a so-called two-stage optimization problem. In the first stage, we make *Here-and-now* decisions before the uncertainty is resolved. In the second stage, we make *Wait-and-see* adjustments after the uncertainty unfolds.

The general formulation of a stochastic program is

$$\min \left(c^T x = \mathbb{E}_\xi [Q(x, \xi)] \right)$$

where $Q(x, \xi)$ optimizes the second-stage costs under the scenario ξ.

Let us take a look at another planning problem. A chemical plant produces two products (A and B) where we have limitations in the raw material availability (max 100 tons/day are shared), and the demand for A is ≤ 40 tons/day and for $B \leq 30$ tons/day. We know that the profit yields \$200/ton for A and \$300/ton for B.

In this case we want to maximize the profit and a deterministic LP looks as follows:

$$\max Z = 200x_A + 300x_B$$

Subject to:

$$x_A + x_B \leq 100$$

$$x_A \leq 40$$

$$x_B \leq 30$$

$$x_A, x_B \geq 0$$

Solving with the Simplex method we find: $x_A = 40$, $x_B = 30$, and $Z = 17.000$.

Suppose now that the demand for B is, with two equally likely scenarios: we might have a **high demand** (50% chance): $x_B \leq 50$ tons/day, or we might have a low demand (50% chance) $x_B \leq 20$ tons/day.

The approach for solving this problem can be as a two-stage stochastic program, where the first stage is to decide on the production *before* the demand is known (in this case) x_A and the second stage is to adjust x_B after the demand is realized.

We have to rewrite our model now:

$$\max Z = 200x_A + 300\left(0.5x_B^{high} + 0.5x_B^{low}\right)$$

subject to:

$$x_A + x_B^{high} \leq 100 \ \ (\text{raw material, high demand})$$

$$x_A + x_B^{low} \leq 100 \ \ (\text{raw material, low demand})$$

$$x_A \leq 40 \ \ (\text{demand for } A \text{ fixed})$$

$$x_B^{high} \leq 50 \ \ (\text{high demand for } B)$$

$$x_B^{low} \leq 20 \ \ (\text{low demand for } B)$$

$$x_A, \ x_B^{high}, x_B^{low} \geq 0$$

The solution now is produce $x_A = 40$ ton/day (first stage) and for the second stage, if the demand for B is high $x_B^{high} = 50$ (profit is \$27,000) And if the demand for B is low: $x_B^{low} = 20$ (profit is \$14,000). The expected profit is \$20,500/day.

Table 8.1 compares our deterministic and stochastic outcomes.

Table 8.1: Key features of the deterministic and stochastic solution.

Aspect	Deterministic LP (Z = $17,000)	Stochastic LP (expected profit = $20,500)
Assumption	Ignores uncertainty (fixed demand for B = 30).	Accounts for demand uncertainty (B = 20 or 50, each 50% likely).
Decision flexibility	Rigid: Commits to x_B = 30 *no matter what*.	Adaptive: Adjusts x_B to 20 (low demand) or 50 (high demand).
Risk	Fails if demand ≠ 30: – If B demand = 50 → Lost profit ($300 × 20 = $6,000). – If B demand = 20 → wastes raw material (10 unused tons).	Hedge risk: – Captures upside (profit $23,000 if high demand). – Minimizes losses ($14,000 if low demand).
Expected profit	$17,000 (only valid if demand = 30).	$20,500 (averages outcomes across scenarios).
Mathematical form	Single LP.	Two-stage program: – First-stage fixes x_A. – Second-stage adjusts x_B.

In stochastic programming, the *expected value of perfect information* (EVPI) quantifies how much you would be willing to pay to eliminate all uncertainty before making decisions. It measures the gap between optimal decisions under uncertainty and optimal decisions if you knew the future perfectly (perfect information).

If you could foresee demand before deciding x_A and x_B we would find for the
– High-demand scenario
  – Optimal plan: x_A = 40, x_B = 50 and Z = $23,000
– Low-demand scenario
  – Optimal plan: x_A = 80, x_B = 20, and Z = $22,000

The expected profit with perfect information is

$$WS = 0.5 \times \$23,000 + 0.5 \times \$23,000 = \$22,500$$

Now, the stochastic solution gave us $20,500, so the EVPI is

$$EVPI = WS - st = \$22,500 - \$20,500 = \$2,000$$

This is the maximum you would pay for a perfect demand forecast. In other words, it is the value of eliminating uncertainty in the demand for B.

We could also calculate the *value of stochastic solution* (VSS), which quantifies the benefits of using a stochastic programming approach over a deterministic one by comparing their expected profits.

Our deterministic solution was $x_A = 40$ and $x_{Bm} = 30$ and the profit was $Z = \$17{,}000$. How would this solution compare to a situation where the demand was high (50). We could have produced $x_B = 50$ and the profit could have been $Z = \$23{,}000$. And how would the outcome be different if we had had a low demand (20). We would still produce $x_B = 30$, but only 20 units could be sold. Suppose we had an unsold inventory cost of $\$50$/ton, the profit would now be: $Z = \$13{,}500$.

The expected profit, deterministic under uncertainty, would now be

$$0.5 \times \$17{,}000 + 0.5 \times \$13{,}500 = \$15{,}250$$

The VSS now is

$$\text{VSS} = \$20{,}500 - \$15{,}250 = \$5{,}250$$

These are the costs of ignoring uncertainty. The VSS measures the value of adapting plans to real-world variability.

8.3.2 Robust Optimization

Robust optimization is a way of finding solutions that are feasible and near-optimal for all realizations of uncertainty, within a predefined set. While stochastic optimization is computing expected performance (which requires probabilities), robust optimization is computing for the worst-case performance (where we do not need probabilities).

A generic robust optimization model (linear) looks like

$$\max c^T x$$

$$\text{s.t.}\, A(\xi)x \leq b \;\; \forall \xi \in U$$

For our planning problem, the robust optimization formulation seeks to maximize profit while ensuring all constraints are satisfied for every possible demand realization within the specified bounds. The worst-case scenario occurs when demand for B is at its minimum (20 tons), as this creates the most restrictive condition for production planning.

The critical constraint involves the uncertain demand for product B. To guarantee feasibility, under all possible demand realizations, we must consider the most restrictive case, in other words: $x_B \leq 20$.

The optimal solution is then $x_A = 40$ and $x_B = 20$ with $Z = \$14{,}000$. We call this a robust optimum or sometimes a fat optimum. This solution possesses three key properties that characterize robust optimization.

First, it ensures all produced quantities can be sold even in the worst-demand scenario ($d_B = 20$ tons). Second, it completely eliminates the risk of unsold inventory. Third, and most importantly, it maintains feasibility for every possible demand realization within the specified uncertainty set.

WURST-CASE SCENARIO

Figure 8.4: PhD candidate Pjotr and the worst-case scenario.

Figure 8.4, PhD candidate Pjotr is taking care of uncertainty in decision-making under uncertainty by means of the "Wurst-Käze" scenario.

8.3.2.1 Comparative Analysis with Stochastic Programming

When contrasted with the stochastic programming approach discussed in Section 8.3.1, fundamental differences emerge in both methodology and outcomes. The stochastic solution, which incorporated probability information (50% chance of high demand at 50 tons and 50% chance of low demand at 20 tons), yielded an expected profit of $20,500 through scenario-adaptive production planning.

The robust solution's guaranteed profit of $14,000 represents the price of robustness – the economic trade-off for absolute protection against worst-case scenarios. This conservative approach proves most valuable when constraint violations carry severe consequences or when probability distributions cannot be reliably estimated.

8.3.2.2 Adjustable Robust Optimization

The basic robust formulation can be enhanced through adjustable robust optimization, which introduces recourse variables that adapt to uncertainty realization. In our production planning context, this would involve:

First-stage decisions: commit to production of A; second-stage decisions: adjust production of B after demand is observed.

The resulting formulation would mathematically resemble the two-stage stochastic program but maintain the robust philosophy of protecting against worst-case scenarios within the uncertainty set.

8.3.2.3 Implementation Considerations

Several practical aspects merit consideration when applying robust optimization to production planning. The uncertainty set [20,50] should reflect true operational limits, potentially estimated from historical demand extremes when available. The price of robustness ($6,500 in this case) must be evaluated against the potential costs of constraint violations.

From a computational perspective, linear robust problems maintain tractability even when scaled to larger problems with multiple uncertain parameters. This computational efficiency, combined with reduced data requirements, makes robust optimization particularly attractive for preliminary design stages or high-risk production scenarios.

8.3.2.4 Extensions and Advanced Formulations

The basic robust framework can be extended in several directions to balance conservatism with performance. The *budget-of-uncertainty* approach introduces a parameter $\Gamma \in [0, 10]$ that controls the degree of conservatism through constraints such as $x_B \leq 30 - \Gamma$. For multiproduct planning problems, the formulation naturally extends to handle shared resources and multiple uncertain demands while maintaining the worst-case protection philosophy.

Robust optimization provides chemical plants with a mathematically sound methodology for production planning under uncertainty that prioritizes absolute protection against worst-case scenarios. While yielding more conservative solutions than stochastic approaches, its strengths lie in guaranteed feasibility, reduced data requirements, and computational tractability. This makes it particularly valuable for high-risk production environments or when probability information is unreliable. The method's rigorous mathematical foundation ensures that solutions will perform acceptably even under the most adverse conditions within the defined uncertainty set.

8.3.3 Chance-Constraint Programming

Chance-constraint programming provides a sophisticated mathematical framework for optimization problems where constraints may be *occasionally* violated within acceptable probability limits. This approach is particularly relevant in chemical engineering applications where strict constraint enforcement may be economically prohibitive or physically unrealistic, yet probabilistic guarantees of feasibility are required.

Consider an optimization problem with an uncertain parameter ξ that follow a probability distribution P. A chance constraint has the form:

$$P(g(x, \xi) \geq 1 - \alpha$$

where $x \in \mathbb{R}$ are the decisions, $g(x, \xi)$ are the system constraints, $\alpha \in \{0,1\}$ is the acceptable violation probability, and $(1 - \alpha)$ is the required confidence level.

For our production planning problem, we might impose:

$$P(x_B \leq d_B) \geq 0.95$$

This indicates that the production of B should meet the demand with 95% probability.

The computational tractability of change constraints depends on their formulation into deterministic equivalents; you could think of Gaussian uncertainty, a scenario approximation, or convex conservative approximations (using Chebyshev or Bernstein inequalities to obtain convex surrogates).

For our planning problem with uncertain demand d_B we reformulate the original constraint $x_B \leq d_B$ to a chance-constrained version: $P(x_B \leq d_B) \geq 0.95$. We model d_B as normally distributed $d_B \sim N(35, 5^2)$ (i.e., the mean is 35 tons, and the standard deviation is 5 tons).

If we standardize $Z \sim N(0,1)$:

$$P \frac{d_B - 35}{5} \geq P \frac{x_B - 35}{5} \geq 0.95$$

Let $Z = \frac{d_B - 35}{5}$:

$$P\left(Z \geq \frac{35}{5}\right) \geq 0.95$$

Using the standard normal CDF (ϕ):

$$P(Z \geq z) = 1 - \phi(z) \geq 0.95 \rightarrow \phi(z) \leq 0.05$$

Now we invert the CFD:

$$z \leq \phi^{-1}(0.05) \approx -1.6449$$

And substitute back:

$$\frac{x_B - 35}{5} \leq -1.6449 \rightarrow x_B \leq 26.78$$

In this case, the optimum is: $x_A = 40$, $x_B = 26.78$, and $Z = \$16.034$. In Table 8.2, the different methods are compared to each other.

Table 8.2: Different approaches for solving an optimization problem with uncertainty.

Method	Production (tons)	Profit	Risk of shortage
Deterministic	$x_A = 40$, $x_B = 30$	$17,000	50% (if $d_B = 50$)
Robust (worst-case)	$x_A = 40$, $x_B = 20$	$14,000	0%
Chance-constrained	$x_A = 40$, $x_B = 26.78$	$16,034	5%
Stochastic	$x_A = 40$, $x_B \in \{20,50\}$	$20,500	Scenario-dependent

8.4 Solution Methods

Optimization under uncertainty demands specialized computational strategies to address the interplay between decision variables and random parameters. This section presents a rigorous treatment of three principal solution paradigms, emphasizing their mathematical foundations and practical implementation in chemical process systems.

8.4.1 Monte Carlo Optimization

Monte Carlo optimization deals with stochastic problems through statistical sampling of uncertain parameters. Given a chance-constrained problem:

$$\min_{x \in X} \mathbb{E}[f(x, \xi)] \text{ s.t. } \mathbb{P}\left(g_j(x, \xi) \leq 0\right) \geq 1 - a_j \ (j-1, \dots)$$

where ξ is a random vector with known distribution, X are deterministic constraints, and a_j are acceptable violation probabilities.

The first step is to generate N independent and identically distributed samples $\{\xi\}_{i=1}^{N}$ from the distribution ξ. There are different methods to create such samples, for example, Latin Hypercube Sampling or Copula methods.

The second step is to create an empirical approximation by reformulating the original problem:

$$\min_{x \in X} \frac{1}{N} \sum_{i}^{N} f\left(x, \xi^i\right) \text{ s.t. } \frac{1}{N} \sum_{i}^{N} \mathbb{I}\left(g_j(x, \backslash x\hat{\imath}) \leq 0\right) \geq 1 - a_j \ \forall j$$

The third step is to apply smoothing techniques, where the nondifferentiable indicator $\mathbb{I}(.)$ can be modeled via a Sigmoid approximation: $\mathbb{I}(z \leq 0) \approx \left(1/\left(1 + e^{kz}\right)\right)$ or via a sample average approximation with mixed integer formulation.

As an example: we want to maximize the expected profit from a batch reactor with
- Uncertainties: $k \sim N(1.2, 0.1^2)$ (rate constant), $\phi \sim U[0.85, 0.95]$ (catalyst efficiency)

- Decision variable: batch time t
- Constraints: 95% probability of conversion $X(t, k, \phi) \geq 0.8$
- The conversion is given as: $X(t, k, \phi) = 1 - e^{-k\phi t}$
- The objective function is: Profit $= 500X - 100t$

Step 1: In Python we can now create 10,000 samples of k and ϕ with:

```
k_samples = np.random.normal(1.2, 0.1, 10,000)
phi_samples = np.random.uniform(0.85, 0.95, 10,000)
```

Step 2: Rewrite the optimization problem:

$$\max_{t \in [1,5]} \frac{1}{10,000} \sum_{i=1}^{10,000} 500X\left(t, k^i, \phi^i\right) - 100t$$

$$s.t. \frac{1}{10,000} \sum_{i=1}^{10,000} \mathbb{I}\left(X\left(t, k^i, \phi^i\right) \geq 0.8\right) \geq 0.95$$

We can solve this model in Python as follows:

```
import numpy as np
from scipy.optimize import minimize

def monte_carlo_optimization(N=10,000):
    # Generate samples
    k = np.random.normal(1.2, 0.1, N)
    phi = np.random.uniform(0.85, 0.95, N)
        # Define empirical objective and constraint
    def objective(t):
        X = 1 - np.exp(-k * phi * t)
        return -(500 * X.mean() - 100 * t)

    def constraint(t):
        X = 1 - np.exp(-k * phi * t)
        return X[X >= 0.8].size / N - 0.95

  # Solve
    res = minimize(objective, x0=3.0,
                   constraints={'type':'ineq', 'fun':constraint},
                   bounds=[(1,5)])
    return res.x
```

We could also follow step 3 and reformulate as a mixed-integer problem:

$$\max_{t,y^i} \frac{1}{10,000} \sum_{i=1}^{10,000} \left(500y^i - 100t\right)$$

$$\text{s.t.} \, y^i \leq X\left(t, k^i, \phi^i\right) + M\left(1 - z^i\right), \quad \sum_{i=1}^{10,000} z\hat{\imath} \geq 9,500$$

where $z^i \in \{0,1\}$ are binary variables and M is a large constant.
The outcomes are:

Batch time (h)	Empirical $P(X \geq 0.8)$	Expected profit ($)
2.5	0.89	1.120
3.2	0.952	1.310
4.0	0.991	1.240

It is noted that the sample size controls the approximation error. Monte Carlo optimization is a power solution method for solving optimization problems with uncertainty, but it is evident that while large sample sizes will reduce approximation error, they will also come at a computational burden.

8.4.2 Decomposition Techniques

Decomposition methods break complex optimization problems into smaller, more manageable subproblems. Here we will discuss the L-shaped method.
Let us consider a two-stage stochastic linear program:
First stage:

$$\min c^T x + \mathbb{E}[Q(x, \xi)]$$

$$\text{s.t.} \, Ax \leq b, \; x \geq 0$$

The second stage (also called the recourse problem):

$$Q(x, \xi) = \min q^T y$$

$$\text{s.t.} \, Wy \leq h(\xi) - T(\xi)x, \; y \geq 0$$

where ξ represents the uncertain parameters.

Step 1: Initialize, start with an initial guess for x (e.g., $x^0 = 0$). Set the lower bound LB $= -\infty$ and upper bound UB $= +\infty$. Set the tolerance to $\epsilon > 0$ (e.g., $= 0.01$).

Step 2: Master problem (outer approximation)

Solve the relaxed version of the first-stage problem:

$$\min c^T x + \theta$$

$$\text{s.t. } Ax \leq b, \ x \geq 0$$

$$\theta \geq \text{optimality cuts (to be added)}$$

The purpose is to approximate the expected recourse $\mathbb{E}(x, \xi)$ using cuts.

Step 3: Subproblems (scenario evaluation): For each scenario ξ^i (e.g., sampled demand values):

$$Q(x^k, \xi) = \min q^T y$$

$$\text{s.t. } Wy \leq h\left(\xi^i\right) - T\left(\xi^i\right)x^k, \ y \geq 0$$

The output of this problem will give the optimal value $Q(x^k, \xi^i)$ and the dual variables π^i (shadow prices).

Step 4: Cut generation: Feasibility cuts, if a subproblem is infeasible, add a cut to exclude x^k. Optimality cuts: approximate $\mathbb{E}[Q(x, \xi)]$ using the subproblem results:

$$\theta \geq \sum_{i=1}^{N} p^i \left[Q(x^k, \xi\hat{\imath}) + (\pi^i)^T T\left(\xi^i\right)(x - x^k) \right]$$

where p^i is the probability of scenario ξ^i.

Step 5: Bounds update. Lower bound: objective value of the master problem and upper bound: beast feasible solution found.

Step 6: Termination: Repeat the procedure until $\text{UB} - \text{LB} \leq \epsilon$.
Consider the following intuitive explanation with a production example. The problem: a factory must decide production levels x before knowing demand ξ After the demand is observed it can adjust production y at a higher cost.

- The master problem
    - Decide initial production x (e.g., produce 100 units)
    - Estimate future costs θ with cuts like: "If you produce 100 units, expect 4,500 in adjustment costs"
- Subproblems
    - Simulate different demand scenarios (e.g., $\xi^1 = 80$ and $\xi^2 = 120$)
    - Calculate costs to adjust production for each scenario:
        - If demand is 80, sell 20 units elsewhere (cost = $200)
        - If demand is 120, buy 20 extra units (cost = $300)

– Cut added to master
  – $\theta \geq 0.5 \times 200 + 0.5 \times 300 +$ adjustment for changing x

This helps the master problem make better decisions in the next iteration.

The nice thing about a composition method is that it will be computationally more efficient as the Monte Carlo optimization because we break down a large problem into smaller pieces. This also allows us to parallelize (solve subproblems simultaneously). The method converges to the optimal solution for convex problems. This chapter has equipped readers with fundamental methodologies to address optimization problems when faced with uncertain parameters in chemical engineering systems. The core approaches – stochastic programming, robust optimization, and chance-constrained programming – each provide distinct strategies to balance performance objectives with the inherent variability of real-world processes.

Stochastic programming emerges as the preferred framework when probabilistic descriptions of uncertainty are available, enabling optimized decisions that account for multiple future scenarios through two-stage or multistage formulations. The L-shaped decomposition method proves particularly valuable here, breaking large-scale problems into tractable subproblems while maintaining convergence guarantees.

For situations where worst-case guarantees are paramount, robust optimization offers a distribution-free alternative, ensuring constraint satisfaction across all possible realizations within defined uncertainty sets. This approach is especially relevant for safety-critical systems, though often at the cost of increased conservatism in solutions.

Chance-constrained programming occupies the middle ground, allowing for quantifiable risk tolerance through probabilistic constraints. This method shines when minor constraint violations are acceptable, provided their likelihood remains below specified thresholds – a common requirement in economic optimization of production systems.

The solution methods discussed, including Monte Carlo sampling and convex reformulations, provide practical computational tools to implement these approaches. A key insight throughout is the fundamental trade-off between solution quality, computational tractability, and risk tolerance that engineers must navigate when selecting an optimization framework.

These techniques find immediate application across chemical engineering domains from plant design and supply chain management to process control and safety analysis. The chapter establishes the foundation for more advanced topics in stochastic optimization while emphasizing the importance of proper uncertainty characterization in decision-making processes. Readers are now prepared to select appropriate methods based on problem requirements, data availability, and risk tolerance – essential skills for modern process optimization challenges.

Further Reading

Birge, J. R., & Louveaux, F. (2011). *Introduction to stochastic programming* (2nd ed.). Springer.

Ben-Tal, A., El Ghaoui, L., & Nemirovski, A. (2009). *Robust optimization*. Princeton University Press.

Ghanem, R., Higdon, D., & Owhadi, H. (Eds.). (2017). *Optimization under uncertainty: Theory and practice*. World Scientific.

Edgar, T. F., & Himmelblau, D. M. (2001). *Chemical engineering process optimization*. McGraw-Hill.

8.5 Exercises

Exercise 1. Deterministic Versus Stochastic Formulation ★

Problem: A plant produces chemical X with profit $50/ton and Y with $80/ton. The total capacity is 100 tons/day. Demand for X is fixed at 40 tons/day, while demand for Y has two equally likely scenarios: 30 tons (50%) or 50 tons (50%).

Tasks:

a)   Formulate the deterministic LP assuming average demand for Y.

b)   Convert to a two-stage stochastic program.

Learning Objective: Distinguish deterministic and stochastic formulations.

Exercise 2. Robust Optimization Basics ★

Problem: Reactor yield depends on uncertain temperature $T \in$ [300, 320] °C. The nominal optimal yield is at $T = 310$ °C, but the yield drops by 2%/°C deviation.

Task: Find the robust operating temperature that maximizes worst-case yield.

Learning Objective: Apply interval-based robust optimization.

Exercise 3. Chance Constraints with Normal Distributions ★★

Problem: Product purity $P \sim N(90, 5^2)$% must satisfy $P \geq 85$% with 95% probability.

Task: Derive the deterministic equivalent constraint.

Learning Objective: Reformulate chance constraints using Gaussian quantiles.

Exercise 4. Monte Carlo Simulation ★★

Problem: Batch reaction time ($^*t^*$) affects yield $Y = 0.5^*(1 - e^{-kt^*})$, where $^*k^* \sim U(0.8, 1.2)$.

Tasks:

a)   Estimate $E[Y]$ for $^*t^* = 2$ h using $N = 1{,}000$ samples.

b)   Find $^*t^*$ to ensure $P(Y \geq 0.4) \geq 0.9$ via sampling.

Learning Objective: Implement Monte Carlo for probabilistic constraints.

Exercise 5. L-Shaped Method ★★★

Problem: Solve the stochastic program:

First-stage: $\min {}^*x^* + \mathbb{E}[Q(x, \xi)]$

Second-stage: $Q(x, \xi) = \min {}^*y^*$ s.t. $^*y^* \geq \xi - {}^*x^*, {}^*y^* \geq 0$

where $\xi \in \{1, 3\}$ with $P(\xi = 1) = 0.6$.
Task:
Perform 2 iterations of the L-shaped method.
Learning Objective: Apply Benders decomposition.

Exercise 6. Multi-objective Robust Design ★★★
Problem: A distillation column has two conflicting objectives:
1. Maximize purity (nominal 95%, ±3% uncertainty)
2. Minimize energy use (nominal 100 kW, ±10 kW).

Task: Formulate as a robust multi-objective problem and sketch the Pareto frontier.
Learning Objective: Trade-off analysis under uncertainty.

Exercise 7. Non-Gaussian Chance Constraints ★★★★
Problem: Waste concentration C follows a lognormal distribution with mean 50 ppm and CV = 0.3. Regulatory limit: $P(C \leq 100 \text{ ppm}) \geq 0.99$.
Task: Derive the deterministic reformulation.
Learning Objective: Handle non-Gaussian distributions.

Exercise 8. Integrated Stochastic-Robust Model ★★★★
Problem: A supply chain has
– Stochastic demands (scenarios ξ_1, ξ_2)
– Robust uncertain transportation costs $^*c^* \in [c_l, c_u]$.

Task: Formulate a hybrid stochastic-robust optimization model.
Learning Objective: Combine uncertainty paradigms.

Group Project: Optimizing a Chemical Supply Chain Under Uncertainty
Duration: 1 week (4–5 team members)
Deliverables: Report + Presentation + Python/Excel Implementation

Project Overview
Teams will design a robust production and distribution plan for a chemical manufacturing company facing:
– **Demand uncertainty** (fluctuating customer orders)
– **Supply uncertainty** (raw material quality variations)
– **Transportation risks** (cost volatility)

Phase 1: Problem Formulation (Days 1–2)
Scenario: A company produces *Chemical A* ($200/ton profit) and *Chemical B* ($300/ton profit) at two plants:
– **Plant 1**: Max 100 tons/day (A + B), 40 tons/day for A
– **Plant 2**: Max 80 tons/day (A + B), 30 tons/day for B

Uncertainties:
1. **Demand**:
   - *A*: 50% chance of 60 tons/day and 50% chance of 40 tons/day
   - *B*: Uniformly distributed between 20 and 50 tons/day
2. **Raw Material Purity**: Impacts yield ± 10% (normally distributed)

Tasks:
1. Formulate as:
   - A deterministic LP (ignoring uncertainty)
   - A two-stage stochastic program
   - A robust optimization model
2. Compare the three approaches theoretically.

Phase 2: Solution Development (Days 3–5)
Team Roles:
1. **Stochastic Lead**: Implement stochastic programming in Python (Pyomo) or Excel Solver.
2. **Robust Lead**: Formulate robust counterpart using uncertainty sets.
3. **Data Analyst**: Generate Monte Carlo samples for yield uncertainty.
4. **Optimization Lead**: Compare solutions and compute:
   - Expected profits
   - Worst-case losses
   - Probability of meeting demand

Methods to Apply:
- **L-Shaped Algorithm** (for stochastic)
- **Ellipsoidal Uncertainty Sets** (for robust)
- **Chance Constraints** (for yield variability)

Phase 3: Analysis and Reporting (Days 6 and 7)
Deliverables:
1. **Technical Report** (10–15 pages):
   - Mathematical formulations
   - Solution code/Excel files
   - Sensitivity analysis (e.g., how profit changes with demand variance)
2. **Presentation** (15 min):
   - Compare solutions visually (e.g., profit distributions)
   - Justify which method is best for the company

9 Optimization in Current and Future Applications

The perfect process doesn't exist, but with optimization, we can inch infinitely closer.

9.1 Introduction

Process optimization is the silent engine driving progress in chemical engineering, from the petrochemical plants that power our economies to the pharmaceutical labs that safeguard our health. In this chapter, we bridge the gap between the *theoretical tools* introduced earlier (linear programming, gradient-based methods, stochastic optimization) and their *real-world applications*, both current and emerging. Here, you'll discover how optimization transcends textbook problems to solve challenges like sustainable energy, smart manufacturing, and even the discovery of new materials.

9.1.1 Why This Chapter Matters?

From Equations to Impact: Optimization isn't just about minimizing costs or maximizing yields; it's about making *informed decisions* under constraints. Whether designing a carbon-neutral reactor or scheduling batches of life-saving drugs, the principles remain the same, but the tools evolve. For example*:* A 2021 study by Dow Chemical used multi-objective optimization to reduce wastewater in a polymer plant by 20% while maintaining production rates, a feat impossible without advanced algorithms (in line with their sustainability goals: Dow's *"2025 Sustainability Targets"* include reducing freshwater intensity in water-stressed regions by 20% (publicly reported in Dow's 2022 ESG Report).

The Digital Transformation: The rise of AI, IoT, and high-performance computing has turned optimization from a *static* design tool into a *dynamic* partner in real-time decision-making. *Case in point:* Digital twins now use optimization to simulate and adjust processes like distillation columns *before* physical changes are made, saving millions in trial-and-error costs.

Future-Proofing Your Skills As quantum computing and autonomous labs emerge, tomorrow's engineers will need to blend classical methods (e.g., simplex algorithms) with cutting-edge tools (e.g., Bayesian optimization). This chapter prepares you for that transition.

In this chapter, we'll explore:
–   **Foundational applications:** Data regression, control, and flowsheeting, where optimization is already routine.

https://doi.org/10.1515/9783111342283-009

- **Advanced frontiers:** AI-driven optimization, probabilistic methods (like Bayesian optimization), and their role in materials discovery.
- **Speculative horizons:** Quantum computing's potential to solve intractable problems (e.g., molecular design).

A Thought Experiment:
Imagine you're tasked with optimizing a solar-powered electrolyzer for green hydrogen production. You'll need to:
- Balance *economic* (cost), *engineering* (efficiency), and *environmental* (carbon footprint) objectives.
- Account for *uncertainties* like weather fluctuations.
- Decide whether to use deterministic models, stochastic programming, or AI surrogates.

This scenario, increasingly common in industry, is where this chapter's tools shine.

9.2 Data Regression

Data regression is a common practice in chemical engineering, often we have experimental data that we want to fit to a model (fitting reaction kinetics, adsorption isotherms, thermodynamic equations of state, and so on).

But we sometimes forget that fitting data to models is actually about solving an optimization problem. We want to minimize the difference between model and data and use the fit parameters as decision variables. A well-known procedure for doing this is the least squares method.

In Figure 9.1, we have plotted experimental data, through this data a third order polynomial has been fitted.

In this example, we have hundred data points and we want to fit the following model to this data:

$$y_p = a_1 + a_2 x + a_3 x^2 + a_4 x^3$$

If we have $N = 100$ data points we could write the model as the product of a matrix and a vector:

$$\begin{bmatrix} y_{p1} \\ y_{p2} \\ y_{p3} \\ \vdots \\ y_{pN} \end{bmatrix} = \begin{bmatrix} 1 & x_1 & x_1^2 & x_1^3 \\ 1 & x_2 & x_2^2 & x_2^3 \\ 1 & x_3 & x_3^2 & x_3^3 \\ \vdots & \vdots & \vdots & \vdots \\ 1 & x_N & x_N^2 & x_N^3 \end{bmatrix} \begin{bmatrix} a_1 \\ a_2 \\ a_3 \\ a_4 \end{bmatrix}$$

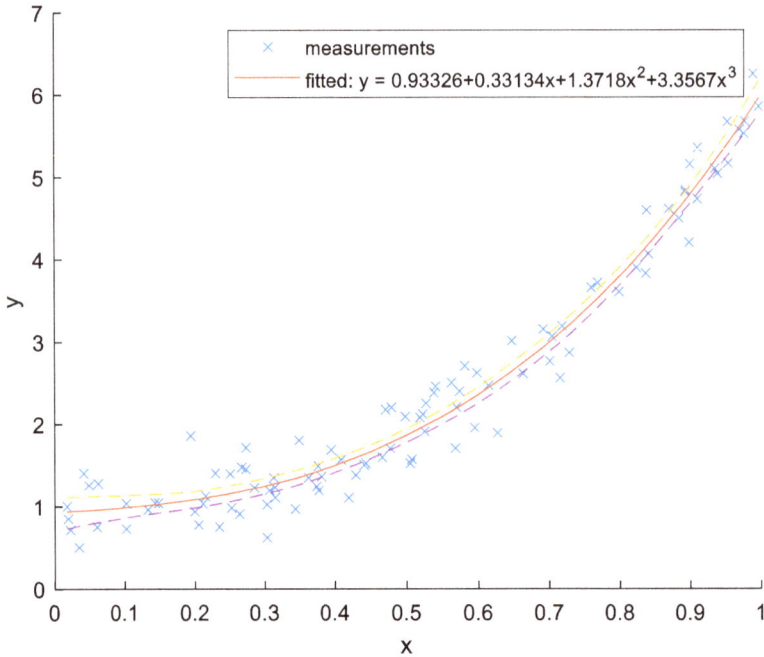

Figure 9.1: Experimental data and fit.

Or more compact:

$$y_P = Xa$$

Here X is called the design matrix.

What can introduce a residual (or error) that is defined as the difference between a data point y_D and a model prediction y_P for each data point n:

$$d_n = (y_{Dn} - y_{Pn})$$

Instead of working directly with the residuals, we can also work with the square of the residual (this is to prevent that large positive residuals cancel out large negative residuals. We want to minimize the sum of squares of the residuals:

$$\min P = \sum_n d_n^2 = \sum_n (y_{Dn} - y_{Pn})^2$$

Because d_n is a vector we can also write the sum as a vector product:

$$\sum_n d_n^2 = d \cdot d = d^T \times d = (y_D - y_P)^T (y_D - y_P)$$

We now much choose the parameter vector a such that the sum of squares of the residuals is minimized; this means the partial derivative with respect to each parameter should be set to zero:

$$\frac{\partial}{\partial a_j}\left[(y_D - Xa)^T(y_D - Xa)\right] = 0$$

The solution to this problem is:

$$a = (X^T X)^{-1} X^T y_D$$

This is the equation that finds the best value for the fit parameters a is called the least squares fit.

For our example, we find $a = [0.9333,\ 0.3313,\ 1.3719,\ 3.3567]$. Overall, the fit (as shown in the figure looks reasonably well. There are several ways of quantifying the goodness of fit. The first thing you could do is to check just for each individual data point how big its residual is. They should randomly move around a value of zero (the smaller the residual, the better). See Figure 9.2.

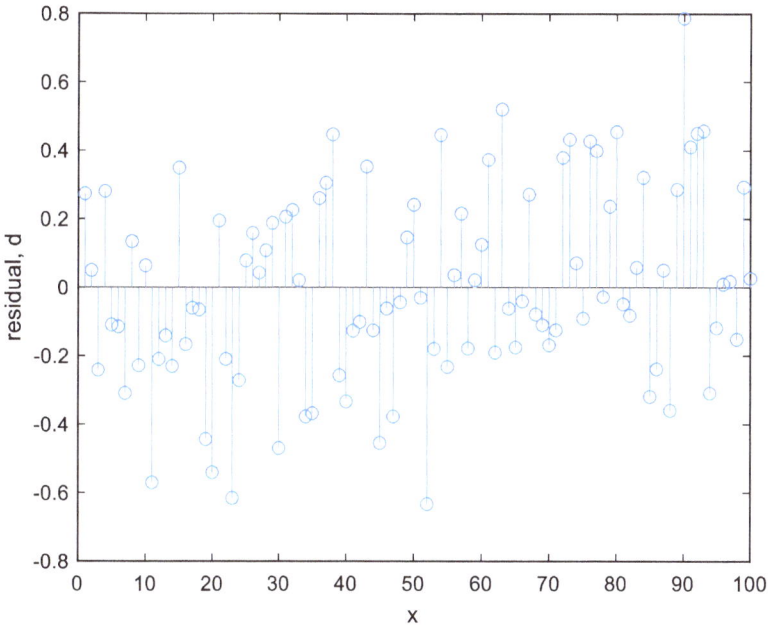

Figure 9.2: A residual plot.

We next can compare variances with each other. We distinguish three types of variances:
The variance measured in the data:

$$\sigma_D^2 = \frac{1}{N} \sum_n (y_D - \overline{y_D})^2$$

where $\overline{y_D}$ is the average of the data points.
The variance of the residuals is:

$$\sigma_R^2 = \frac{1}{N} \sum_n (d_n)^2$$

and the variance of the model is:

$$\sigma_P^2 = \frac{1}{N} \sum_n (y_P - \overline{y_P})^2$$

where $\overline{y_P}$ is the average of the model predictions. If the error is uncorrelated we can state that:

$$\sigma_D^2 = \sigma_R^2 + \sigma_P^2$$

And the correlation coefficient is:

$$R^2 = \frac{\sigma_P^2}{\sigma_D^2} = 1 - \frac{\sigma_R^2}{\sigma_D^2}$$

In our case $R^2 = 0.9652$ The closer the correlation coefficient is to unity, the better the fit (model and data are more strongly correlated). Programming environments such as MATLAB (e.g. cftool, fit, and nonlinfit) and Python (e.g. curve_fit) have many libraries for specifically designed for curve fitting.

9.3 Process Control

9.3.1 Optimal Control

Optimal control is also called "dynamic optimization" or sometimes "the calculus of variations". The basic ideas were developed in the 1940s (by Lev Pontryagin) and 1950s (by Richard Bellman). The concepts have really changed the world, from auto pilots in airplanes to self-driving cars nowadays. Also in chemical engineering optimal control is of utmost importance. In optimal control we are using optimization concepts to find the best decisions to maximize or minimize an objective, but we will predict how the decisions have to move over a time trajectory.

Illustrative example of a reactor

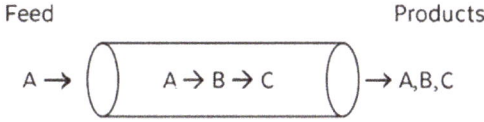

Feed   Products

A → (| $A \to B \to C$ |) → A,B,C

$$\frac{dx_A}{dt} = -k'_A x_A \exp(-E'_A / RT)$$

$$\frac{dx_B}{dt} = -k'_A x_B \exp(-E'_B / RT) - k'_B x_S \exp(-E'_B / RT)$$

$$\max_{T(t)} F = x_B(t_F)$$

We want o determine the optimal
temperature profile as to
maximize the concentration of
component B at the outlet.

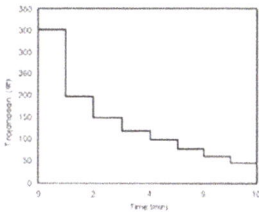

Common sense: High temperature at the beginning (to ma:

Figure 9.3: Optimal control for a tubular reactor.

As an example (see Figure 9.3), suppose we have a tubular reactor that converts a chemical A into B, but B can also further react in to C. We know that the reaction rate from A to B and from B to C depend on the temperature in the reactor. We are interested to maximize the production of B. This means to rise the temperature in the reactor (because A will go to B), but if we rise the temperature too much, B will further react into C. We should actually define a temperature profile over the time. Most likely we will heat up the reactor (to make B) and then decrease (to prevent production of C). The question: how do we choose the right profile of temperature over the time is answered by doing dynamic optimization (or optimal control). So how can we find such profiles?

We have several possibilities: via the so-called maximum principle (on the basis of Pontryagin's ideas), we could use dynamic programming (on the basis of Bellman's ideas) or we could use direct or parametrization methods.

The maximum principle starts of with formulating the optimization problem in the Bolzano form:

$$\max V(T) = \sum_i c_i C_i(T)$$

$$\text{s.t.} \frac{dX_i}{dt} = f(X_1, .., X_n, u_1, ...u_m, t)$$

where X are state variables, u are control variables and f are stirring functions. In analogy to our tubular reactor problem, the concentrations of A, B, and C are state variables, the temperature is the control variable and the component balances are the stirring functions.

The Bolzano form occurs basically everywhere in economics, in biology and in engineering as well.

The maximum principle looks in a way similar to the Lagrange multiplier method, we start off with defining a Hamiltonian function:

$$\mathcal{H} = \sum_i \phi_i f_i$$

Here ϕ_i are called adjoint variables (similar like Lagrange multipliers). We now need to find controls (u) that ensure that:

$$\frac{\partial \mathcal{H}}{\partial u_k} = 0$$

In order to find an optimum. We can find the adjoint variables from:

$$\frac{d\phi_i}{dt} = -\sum_i \frac{\partial f_i}{\partial X_i} = -\frac{\partial \mathcal{H}}{\partial X_i}, \quad \phi_i(T) = c_i(T)$$

Let us try out the maximum principle with an example from economics, a consumption problem.

Suppose we have the following problem:

$$\max U(T) = \int_0^T \ln c(t) e^{-at} dt$$

$$\text{s.t.} \frac{dA}{dt} = u(t) + rA(t) - c(t), \quad A(t) \geq 0$$

In this problem $U(t)$ is the utility, $A(t)$ are the financial assets, $c(t)$ is the consumption profile and $y(t)$ is the income. Effectively, we have income and assets and we want to find a consumption profile that maximizes our financial happiness (= utility).

The format above is not yet in the Bolzano form, we can rewrite the model:

$$\max V(T) = U(T) + \lambda A(T)$$

$$\text{s.t.} \frac{dU}{dt} = \ln c(t) e^{-at}$$

$$\frac{dA}{dt} = y(t) + rA(t) - c(t)$$

With the Bolzano form in place we can define the Hamiltonian:

$$\mathcal{H} = \phi_U \left\{ \ln c(t) e^{-at} \right\} + \phi_A \left\{ y(t) - rA(t) - c(t) \right\}$$

We now have to find the $c(t)$ that sets the derivative of the Hamiltonian to zero:

$$\frac{\partial \mathcal{H}}{\partial c} = \frac{\phi_U e^{-at}}{c} - \phi_A = 0 \Rightarrow c = \frac{\phi_U}{\phi_A} e^{-at}$$

Now we need to find the adjoint variables:

$$\frac{d\phi_U}{dt} = -\frac{\partial \mathcal{H}}{\partial U}, \quad \phi_U(T) = 1$$

which gives:

$$\phi_U(t) = 1$$

and

$$\frac{d\phi_A}{dt} = -\frac{\partial \mathcal{H}}{\partial A} = -\phi_A r, \quad \phi_A(T) = \lambda$$

which gives:

$$\phi_A(t) = \lambda e^{r(T-t)}$$

Substituting the adjoint variables in the optimized control gives us the dynamic consumption profile that maximizes the utility:

$$c(t) = \frac{e^{-at}}{\lambda e^{r(T-t)}}$$

The maximum principle gives an analytical solution, but it is (computationally) very expensive for large systems. In some cases, we could use Bang-bang control, as an outcome of Pontryagin's principle, which effectively means to switch from minimum to maximum at given times.

Bellman's Principle of Optimality states that an optimal policy has the property that, regardless of initial decisions, the remaining decisions must form an optimal policy for the state resulting from those initial decisions. In dynamic optimization, where decisions are made sequentially over time or stages, this principle allows complex problems to be broken into simpler subproblems.

For example, consider optimizing a **batch reactor's temperature profile** over time to maximize yield while minimizing energy use. Using Bellman's approach, we divide the process into discrete time intervals (stages). At each stage, we solve for the optimal temperature decision *backwards* (from final to initial time), storing intermediate solutions. This avoids brute-force computation by reusing prior results (memorization), reducing exponential complexity to polynomial time.

The main steps are:
1. **Decomposition:** Split the problem into stages (e.g., reaction time intervals).
2. **Recursive Backward Solution:** Solve the final stage first, then iteratively determine optimal decisions for preceding stages using Bellman's equation:

$$V_i(x_t) = \min_{u_t}\{g_i(x_t, u_t) + V_t + 1\ (f_{t(x_t, u_t)})\}$$

where V_t is the value function (optimal cost-to-go), x_t is the state (e.g., concentration), u_t is the control (e.g., temperature), and g_t is the stage cost.

We could also use direct methods, where we replace the integral by a summation and the differential equations with finite differences:

$$\min F = \sum_k |\sigma(x_k, u_k)\Delta t + \zeta(x_k, T)$$

$$\text{s.t } x_k = f(x_{k-1}, u_k)\Delta t + x_{k-1}$$

This problem can be solved as an NLP. The quality of this solution strongly depends on how small you take Δt.

Another common way to find optimal profiles for the controls in dynamic optimization is via vector parametrization. In parametrization, a polynomial (or other form) for the control is assumed and added as a constraint to the problem:

$$u(t) = a - 0 + a_1 t + a_2 t^2 + \cdots + a_n t^n$$

The optimal values for the parameters a have no to be computed. For our reactor example: we could assume a second-order polynomial or the temperature over the time, and optimize the coefficients in such way that the yield for B is maximized.

Optimal control can be used for many different types of problems, for example in the periodic cleaning of ultra filtration membranes used for purifying surface water, or for the optimal grade switching in reactive distillation for the production of polyesters/polyamides, or for the optimal trajectories for a simulated moving bed system that produces fructose from glucose.

In optimal control, there are a number of developments. Multiple shooting techniques allow for reformulations (say from NLP's to SQPs) to reduce numerical complexity. Collocation methods can be used to deal with stability and numerical issues. And new discretization techniques can deal with high index constraints and singular arcs.

9.3.2 Control System Design

Process control is concerned with suppressing disturbances in process systems. These disturbances can come from variation in feed, changes in performance of equipment, tear and wear, weather influences, and so on. A control system ensures that process systems are kept at the desired operating conditions. As an example: typically level, flow, pressure, temperature are controlled in such way that processes are safe, economically viable, that product specs are met or that equipment is protected. Process control is matured and besides devices (pumps, valves, sensors, etc.) also algorithmic methods are at hand to keep process conditions steady. We name feedback control, feedforward control, cascade control, split range control, dead-time compensators, decouplers and more advanced distributed and model predictive control methods. In Figure 9.4, a block diagram with a feedback loop is shown.

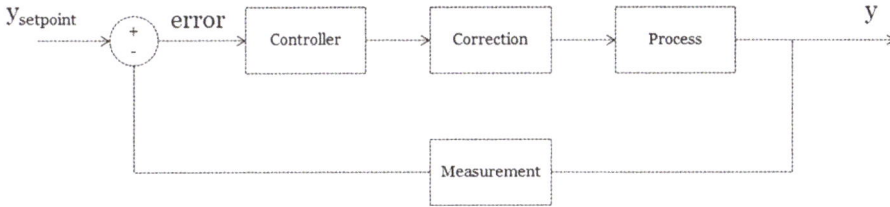

Figure 9.4: Feedback control loop.

A process has a certain output. For example, the liquid level in a tank, denoted with y the output is compared to a desired value of the tank level, the setpoint ($y_{setpoint}$). The difference between output and setpoint is called the error. Now a feedback controller calculated, based on the size of the error, how much of a corrective action (say opening or closing a valve) is required to bring the actual tank level closer to its setpoint.

We often do not think about it, but the feedback controller is doing optimization, it tries to minimize the error. A well-known concept to compute how much corrective action is needed is called the PID-controller law:

$$u(t) = K_p e(t) + K_i \int_0^t e(\tau)d\tau + K_d \frac{de(t)}{dt}$$

Here $u(t)$ is the corrective action and $e(t)$ the error. The constants $K_p, K_i,$ and K_d are the proportional-, integral, and derivative tuning parameters. By tuning these parameters the correction is based on the actual error, its integral and/or its derivative.

The PID controller equation can actually be derived from minimizing a quadratic cost function:

$$J = \int_0^\infty \left[e^2(t) + \rho u^2(t) \right] dt$$

By minimizing this cost function as a dynamic optimization problem we will reduce error and prevent that the control action is too drastic.

Actually we can also tune the controller via optimization. Modern auto-tuners actually solve:

$$\min_{K_p\ K_i\ K_d} \int_0^T w_1 e^2(t) + w_2 u^2(t)dt$$

With weights w_1 and w_2 prioritizing setpoint tracking versus actuator wear.

Beyond PID we nowadays have advanced controllers (e.g. model predictive control: MPC) which explicitly solve optimization problems like:

$$\min_u \sum_k^N ||y(k) - y_{setpoint}||^2 + \lambda||\Delta U(k)||^2$$

Subject to process constraints. In MPC, we use a (linear) model to predict a few time steps a head what the process output y does and then use a least squares optimization to find the optimal trajectory for the input u this procedure continuously repeats where the prediction time horizon rolls forward.

From the Expert: **Prof. Sigurd Skogestad**
Norwegian University of Science and Technology, Trondheim, Norway

Optimization Using Feedback Control

Did you know that optimization problems can be solved online using feedback control? Consider the task of minimizing a scalar cost function J, or equivalently, maximizing the profit $-J$. Typically, J represents an economic quantity with units such as [$/s]. We assume that we can influence J through a manipulated input variable u, and that we may also have access to measurements y. Furthermore, we assume the system is static (i.e., without dynamics), so we can write $J = J(u)$.

Our objective is to minimize $J(u)$ in real time using feedback control. There are three main approaches to doing this: purely data-based, model-based, and SOC (offline model based). In addition, there are hybrid methods and combinations. Actually, all three methods can be combined.

1 Purely Data-Based Feedback Optimization (ESC)

Also known as extremum seeking control (ESC), hill-climbing, greedy search, or perturb-and-observe.

This method relies entirely on measurements of the cost J (which may not always be available) and does not use a process model. It is simple, but generally slow, especially when gradient estimation is involved. In classical ESC, sinusoidal perturbations are used to estimate the gradient, but this is not always an efficient method.

A simpler and often more intuitive version is the **perturb-and-observe** algorithm (commonly used, for example, to optimize the position of wind turbines):

Step 1: Apply a perturbation Δu to the input:
$u(k) = u(k - 1) + \text{sign}(k) \, \Delta u$
where k is the current time sample and sign(k) alternates between +1 and −1.

Step 2: Observe the cost $J(k + 1)$. If $J(k + 1) < J(k)$, then the system is moving in the right direction so keep the same sign; otherwise, reverse the direction (change sign(k)). Repeat from Step 1.

This method has two main tuning parameters:

The **step size** Δu: Larger steps speed up convergence but cause larger oscillations around the optimum.

The **sampling time** T_s: This must be long enough to let the system respond. A rule of thumb is $T_s \approx 3\tau$, where τ is the system's time constant.

This method can be combined with faster model-based optimization methods like RTO by letting u be the **bias** for the RTO-gradient. This hybrid approach is often called **modifier adaptation**.

2 Standard Model-Based Optimization (RTO)

Real-time optimization (RTO) uses a detailed nonlinear model and does not require that the cost J is measured.

The system model is used to compute the input $u(k)$ that minimizes the cost J. Measurements y are used online to update selected model parameters (e.g., efficiencies). RTO minimizes the cost J while explicitly handling constraints.

3 Self-Optimizing Control (SOC)

Here, the model is used offline to find good "self-optimizing" variables c. The online implementation uses a simple PID controller to keep c and its setpoint c_s.

The goal of SOC is to find a **"magic variable"** c to control – ideally one where a constant setpoint c_s leads to near-optimal performance despite disturbances. This allows the optimization to be embedded into the fast control layer.

In its simplest form, $c = y$ is a single measurement. The variable c is chosen such that its setpoint c_s is **insensitive to disturbances**, yet c is **sensitive to input changes** (i.e., has a large gain from u to c).

More generally, we can use combinations of measurements: $c = Hy$ where H is a matrix. Ideally, this corresponds to the **gradient**: $c = J_u$. See the paper by Bernardino and Skogestad (2024) for methods to estimate this gradient.

3C Self-Optimizing Control with Constraints

SOC can be extended to handle constraints by incorporating Lagrange multipliers into the control strategy. In this approach, the **multiplier acts as a manipulated variable in an upper slow control layer**. Constraint violation can be avoided using **override logic** (Dirza and Skogestad, 2024). In essence, this is a clever trick where a PI controller is used to iteratively solve a set of equations numerically.

Combinations

These approaches can be combined:

1 + 2: ESC (slow) sends bias for gradient J_u (modifier adaptation) to RTO. The bias means that the RTO-gradient will not be zero. This is to correct for model errors, unmeasured disturbances and measurement errors in the RTO-layer.

1 + 3: ESC (slow) updates setpoints to SOC (fast).

2 + 3: RTO (slow) updates setpoints to SOC (fast).

1 + 2 + 3: ESC (slow) sends bias for J_u to RTO (faster), which updates setpoints to the SOC layer (fastest).

References

[1] R. Dirza and S. Skogestad, Primal-dual feedback-optimizing control with override for real-time optimization, Journal of Process Control 138, 103208 (2024)

[2] L.F. Bernardino and S. Skogestad, Optimal measurement-base cost gradient estimate for feedback real-time optimization, Computers & Chemical Engineering, 108815 (2024).

9.4 Process Simulation and Flowsheet Optimization

Process simulators like Aspen Plus, CHEMCAD, and gPROMS play a central role in chemical engineering by solving large systems of nonlinear equations that describe material and energy balances, phase equilibria, and reaction kinetics. These tools rely heavily on numerical optimization techniques to converge flowsheets, handle recycles, and optimize process performance.

9.4.1 Equation-Oriented Versus Sequential Modular Approaches

Modern simulators use two primary strategies to solve flowsheets. The equation-oriented (EO) approach formulates the entire flowsheet as a single system of nonlinear equations and solves those simultaneously using Newton-type methods. This method is computationally efficient for large systems but requires careful initialization. In contrast, the sequential modular (SM) approach solves unit operations one at a time, iterating around recycle loops until convergence. While more robust for initial guesses, SM can be slower due to nested iterations. In Figure 9.5, a modular flowsheet simulator (AspenPlus) is shown.

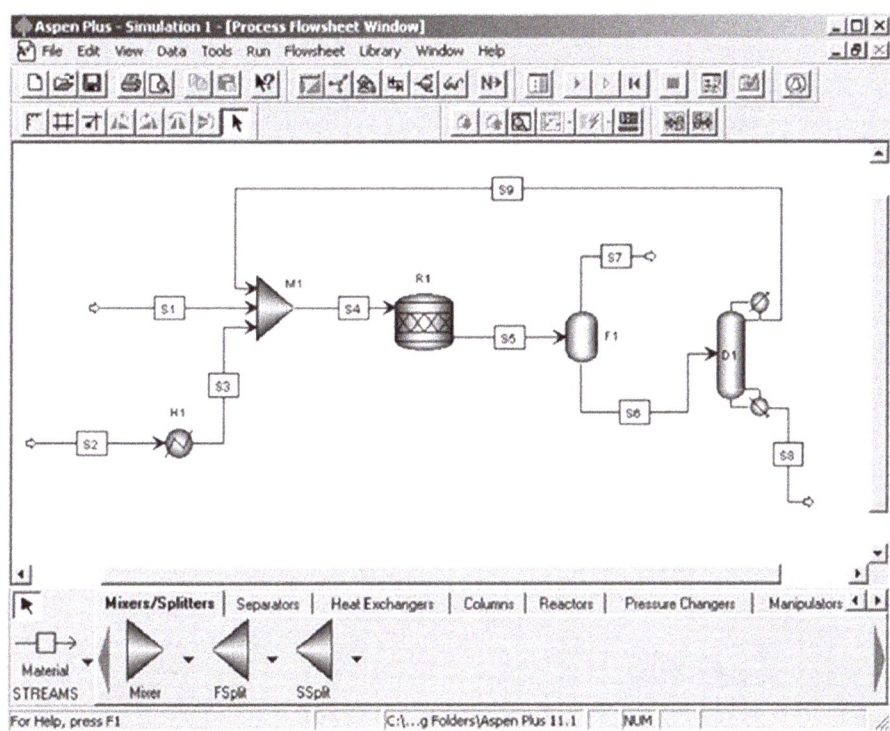

Figure 9.5: Screenshot of the AspenPlus interface.

9.4.2 Handling Recycles and Tear Streams

Recycle streams introduce algebraic loops that require iterative solutions. Simulators uses tear stream algorithms to break these loops, selecting streams where initial guesses are updated iteratively. Techniques like the Wegstein acceleration or Newton-Raphson method are applied to minimize the number of iterations. For example, in a methanol synthesis loop with unreacted gas recycle, the simulator may tear the recycle stream and iteratively adjust its composition and flow rate until the loop converges within a specified tolerance.

9.4.3 Design Specifications and Sensitivity Analysis

Process simulators allow users to impose design specifications, which are constraints that must be satisfied by adjusting decision variables. For instance, a distillation column might require a 95% purity specification on the distillate, achieved by varying the reflux ratio. Internally, the simulator solves this as an optimization problem, often using gradient-based methods to minimize the difference between the actual and target values.

Sensitivity analysis tools complement this by quantifying how process variables (e.g., reactor temperature and feed ratio) affect key outputs (e.g., yield and energy consumption). This is particularly useful for identifying optimal operating windows or troubleshooting bottlenecks.

9.4.4 Flowsheet Optimization

Beyond steady-state simulation, flowsheet optimization involves adjusting decision variables (e.g., operating condition and, equipment sizes) to minimize costs or maximize performance. Simulators leverage nonlinear programming (NLP) or sequential quadratic programming (SQP) to solve these problems. For example, optimizing a heat exchanger network might involve minimizing total annualized cost (capital + operating) by varying heat exchanger areas and utility loads, subject to temperature constraints.

9.4.5 Challenges and Numerical Considerations

Convergence difficulties often arise from poor initialization, stiff equation systems, or discontinuous physical properties (e.g., phase changes). Simulators address these by:
–   Providing robust initialization routines
–   Using homotopy continuation to guide solutions
–   Implementing automatic scaling of variables

9.4.6 Future Directions

Emerging trends include hybrid EO/SM approaches for faster convergence, integration with machine learning for property prediction, and cloud-based optimization for large-scale problems. These advances continue to expand the role of process simulators in designing sustainable and efficient chemical processes.

By combining rigorous equation-solving with optimization algorithms, process simulators provide an indispensable tool for flowsheet development, enabling engineers to explore complex process alternatives efficiently.

9.5 Batch Scheduling

9.5.1 Continuous Versus Batch Processing: Key Considerations

Continuous processes are the backbone of industries producing large-scale commodities such as petroleum, plastics, and paper, where efficiency and low production costs are paramount. However, batch processing is indispensable for specialized products like pharmaceuticals, specialty chemicals, and food products, where production volumes are smaller, and customization is critical. Batch processes offer flexibility, making them ideal for scenarios involving low flow rates, intermittent demand, or hazardous materials that require stringent safety controls. Additionally, batch processing is well-suited for multi-product facilities where the same equipment is used to manufacture different products.

One of the primary challenges in batch processing is determining the optimal vessel size, which balances capacity and cost. Another critical factor is batch time optimization, where the goal is to minimize production time while meeting quality constraints. Dynamic control is also essential, particularly for exothermic or sensitive reactions, where parameters such as temperature and flow rates must be carefully adjusted.

A classic example is the exothermic batch reactor, where a reaction of the form $n_1 A \rightarrow n_2 B$ occurs. The optimal temperature profile that maximizes conversion while minimizing batch time is given by:

$$T(t) = \frac{E_2 - E_1}{R \ln\left(\frac{c_{A_2}^n k_2 E_2}{c_{A_1}^n k_1 E_1}\right)}$$

plotting $T(t)$ reveals an exponential decay, which helps determine the shortest feasible batch time.

9.5.2 Batch Scheduling: Fundamentals and Strategies

Batch scheduling is a decision-making process that allocates resources, such as equipment and labor, to tasks over time. This is particularly crucial in industries with multi-product campaigns, where the sequencing of batches significantly impacts efficiency. The primary objectives of scheduling include minimizing the make span (total production time), maximizing throughput or profit, and reducing idle times and changeover costs.

Two common production strategies are single-product campaigns (SPCs) and mixed-product campaigns (MPCs). In SPCs, all batches of one product are completed before switching to another, which reduces changeovers but may lead to higher inventory costs. In contrast, MPCs alternate batches of different products, which can reduce cycle time but may require more frequent cleanups. For example, in a plant producing products A (with stage times of 5, 1, and 1 h) and B (1, 2, and 2 h), an SPC results in a make span of 24 h, whereas an MPC reduces it to 18 h, though additional cleaning may be necessary. Figure 9.6 shows Gantt chart for different campaigns.

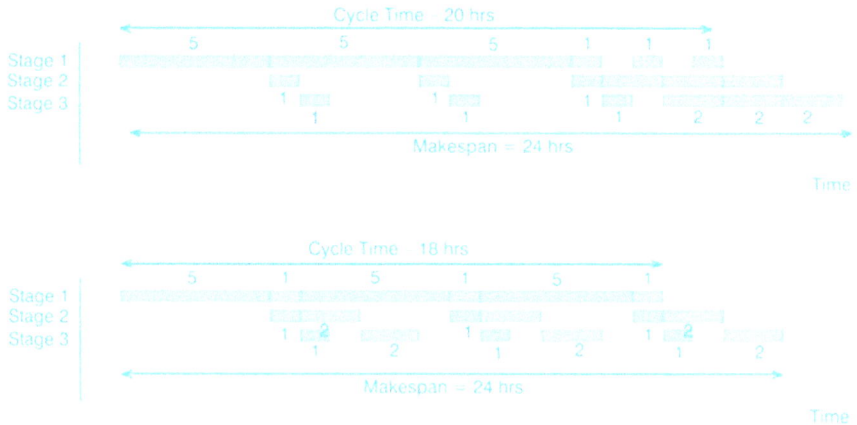

Figure 9.6: Schedules for single- and mixed-product campaigns. Light bars for product A, dark bars for product B.

Plant configurations also play a role in scheduling. Flow shop plants require all products to follow the same sequence, whereas job shop plants allow different products to undergo different stages or sequences. Transfer policies further influence efficiency: a zero-wait policy enforces immediate transfer between stages, which is restrictive and leads to longer cycle times; unlimited intermediate storage provides flexibility but requires additional capacity; and no intermediate storage keeps batches in their equipment until the next stage is ready.

Efficiency can be improved through strategies such as adding parallel units, which reduce cycle time,for instance, introducing a second bioreactor can halve the processing time from 12 to 6 h. Intermediate storage can also decouple stages, allowing them to operate independently.

9.5.3 Advanced Scheduling with State-Task Networks and Mathematical Optimization

State-task networks (STNs) provide a structured way to represent batch processes by distinguishing between states (materials such as feeds, intermediates, and products) and tasks (operations that transform materials). Unlike traditional recipe networks, STNs eliminate ambiguities in material flow. For example, a task may produce a single intermediate that is shared by subsequent tasks, or it may require inputs from multiple sources, allowing for flexible feedstock usage. Figure 9.7 shows a recipe network with its ambiguity and possible STNs.

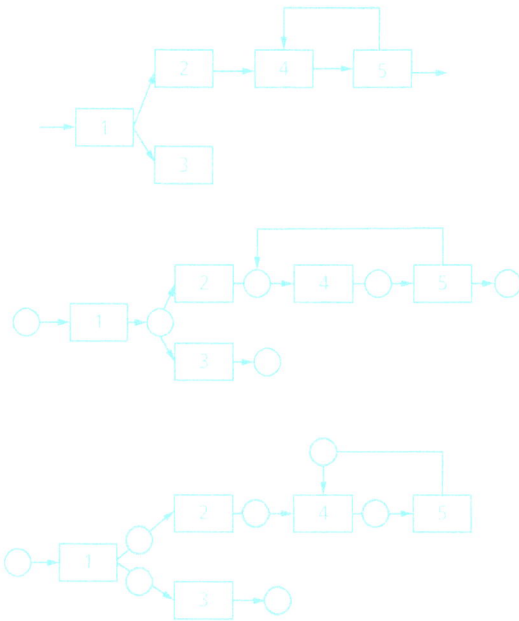

Figure 9.7: Above: A recipe network that might have ambiguities. In the middle and below: Two different state task networks that distinguish between a state (sphere) and a task (square).

To optimize scheduling, mixed-integer linear programming (MILP) is widely used. The formulation includes binary variables W_{ijt} indicating whether task i starts on unit j at time t continuous variables B_{ijt} representing batch sizes, and S_{st} tracking storage levels. Key constraints ensure that units are not over-allocated, mass balances are maintained, and demand is met. The objective function typically maximizes profit by balancing sales, costs, and resource utilization.

A practical example is an ice cream factory with two stages: mixing and packing. The mixing line produces four varieties, while two packing lines handle different products. Constraints include 4-hour changeovers between products and varying aging times (4–12 h depending on the product). Without considering aging, the optimal make span is 104 h, but incorporating aging extends it to 112 h, demonstrating how additional constraints impact scheduling.

9.5.4 Practical Implementation and Tools

Implementing batch scheduling models requires specialized software such as AIMMS, GAMS, or Python-based tools like Pyomo. The process involves defining sets (tasks, units, and time periods), inputting parameters (processing times and batch sizes), formulating constraints and objectives, and solving the model to generate schedules, often visualized using Gantt charts. In summary, batch processes are essential for high-value, low-volume products, and effective scheduling, through STNs and MILP, optimizes resource use, reduces costs, and ensures timely production.

9.6 Enterprise-Wide Optimization

Enterprise-wide optimization (EWO) represents a critical intersection between chemical engineering and operations research, focusing on the holistic improvement of supply chains, manufacturing processes, and distribution networks. Unlike traditional supply chain management, which emphasizes logistics through linear models, EWO places greater emphasis on manufacturing facilities, often requiring sophisticated nonlinear process models. This approach is particularly valuable in industries like fast-moving consumer goods (FMCG), where companies manage thousands of stock-keeping units (SKUs) with varying production requirements. The core of EWO lies in optimizing three key operational layers: scheduling, which deals with short-term resource allocation; planning, which addresses medium-term procurement and distribution strategies; and real-time optimization, which enables dynamic adjustments during operations.

9.6.1 Scheduling and Planning in EWO

Scheduling plays a pivotal role in manufacturing by determining how limited resources are allocated to tasks to achieve objectives such as minimizing production time or costs. While manual scheduling methods like spreadsheets are still used, they often fall short in complex scenarios. Advanced techniques, particularly mixed-integer linear programming (MILP), have proven effective, improving production capacity utilization by 10–30%. Three primary scheduling approaches are commonly employed. Discrete-time models divide the production timeline into fixed intervals, ensuring precision but often resulting in large, computationally intensive models. Continuous-time models, on the other hand, allow for variable time slots, reducing complexity but introducing challenges in determining optimal slot lengths. Precedence-based models focus on sequencing tasks, making them ideal for operations with frequent changeovers, though they struggle with scalability as batch numbers increase.

A practical example of scheduling can be seen in ice cream production, where manufacturers must balance changeover times (typically 4 h) with product-specific aging requirements (ranging from 4 to 12 h). Using MILP, optimal schedules can be developed, though constraints such as aging can extend the total production time significantly.

Beyond scheduling, tactical planning addresses broader supply chain decisions over longer horizons, often spanning a year. Unlike scheduling, planning aggregates data, such as grouping products into families, and typically overlooks sequence dependencies to simplify the model. Figure 9.8 shows how planning and scheduling can be integrated in a pyramid.

Modern advancements in planning include integrated MILP models that synchronize production and distribution, outperforming traditional decoupled approaches. Multi-echelon inventory systems have also gained traction, optimizing safety stock levels across supply chains and yielding substantial cost savings, as demonstrated by Procter & Gamble's $1.5 billion reduction in inventory costs.

Figure 9.8: Temporal layers in enterprise-wide optimization.

9.6.2 Optimization Techniques and Emerging Challenges

Mixed-integer linear programming (MILP) has become a cornerstone of EWO due to its ability to model complex decision-making processes with both continuous and binary variables. The branch-and-bound algorithm is a widely used method for solving MILP problems, operating by solving relaxed linear programming models, branching on fractional variables, and pruning suboptimal paths. Advances in solver technology, such as those seen in CPLEX and GUROBI, have dramatically improved computational efficiency, with some applications running 100 million times faster than two decades ago. Despite these advancements, the NP-complete nature of MILP means that problem complexity grows exponentially with scale, posing ongoing challenges.

Three major challenges dominate EWO today. The first is flexibility, particularly in handling uncertainty. Stochastic models address this by incorporating probability distributions for variables like demand fluctuations, enabling more resilient decision-making. The second challenge is sustainability, where multi-objective optimization techniques like Pareto frontiers help balance economic performance with environmental impact. For instance, an ice cream supply chain was able to reduce its environmental footprint by 12% without incurring additional costs by using the ε-constraint method. The third challenge is complexity, as large-scale, nonlinear models require innovative solutions such as decomposition, reformulation, or parallel computing to remain tractable.

9.6.3 Future Directions and Strategic Implications

Looking ahead, EWO is poised to integrate increasingly sophisticated tools for multi-criterion decision-making, uncertainty management, and reverse engineering. These advancements will enable companies to design supply chains that are not only cost-effective but also sustainable and resilient. The ability to optimize across multiple objectives, such as minimizing costs while reducing carbon emissions, will be a key differentiator in competitive markets; you can think of for example smart grid developments (Figure 9.9).

In summary, EWO transforms traditional decision-making by applying advanced optimization techniques to entire supply chains. While computational challenges remain, ongoing innovations in algorithms and hardware continue to expand the boundaries of what is possible, offering companies new ways to enhance efficiency, sustainability, and profitability.

9.7 Bayesian and Probabilistic Optimization

Bayesian optimization (BO) is a powerful tool for optimizing expensive, noisy, or poorly understood systems, common challenges in chemical engineering. Unlike trial-

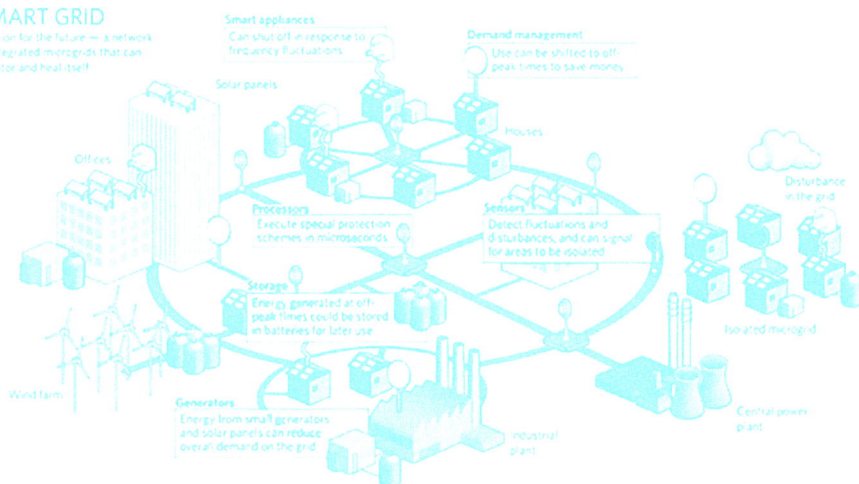

Figure 9.9: Smart grid developments.

and-error approaches, BO uses probability to strategically guide experiments toward optimal conditions.

BO iteratively builds a statistical model (typically a Gaussian process) of the process response surface. After each experiment, it updates the model and calculates an "acquisition function" to determine the most informative next test. For example, when optimizing a reaction yield Y with temperature T (50–150 °C) and catalyst concentration C (0.1–1.0 mol/L), BO might follow this sequence:

1. **Initial data**: 5 random (T,C) experiments yielding 60–80% yield.
2. **Model fitting**: A Gaussian process predicts Y across the parameter space, with uncertainty bounds.
3. **Next experiment**: The algorithm selects (112 °C, 0.63 mol/L) (112 °C, 0.63 mol/L) as the most promising point.
4. **Convergence**: After 10–15 iterations, it identifies $T = 118$ °C, $T = 118$ °C, $C = 0.58$ mol/L, $C = 0.58$ mol/L as optimal, achieving 92% yield.

9.7.1 Why Chemical Engineers Use BO

- **Efficiency**: Reduces experiments by 50–80% compared to grid search.
- **Noise tolerance**: Handles measurement variability better than gradient-based methods.
- **Black-box optimization**: Works without mechanistic equations (e.g., optimizing neural network-controlled processes).

Implementation Example

Using Python's **scikit-optimize** to maximize a simulated reaction yield:

```python
from skopt import gp_minimize
import numpy as np

def reaction_yield(params):
    T, C = params
    # Remove negative sign and reduce noise
    return 0.8 - 0.001*(T-100)**2 - 0.2*(C-0.5)**2 + np.random.normal
(0, 0.01) # Small noise
result = gp_minimize(
    lambda x: -reaction_yield(x),  # Negative for maximization
    [(50, 150), (0.1, 1.0)],
    n_calls=20,
    random_state=42
)

print(f"Optimal: {int(result.x[0])}°C, {result.x[1]:.3f} mol/L |
Yield: {-result.fun:.2f}")
```

Output: Optimal: 101 °C, 0.536 mol/L | Yield: 0.81.

Key Applications

- **Process intensification**: Optimizing operating conditions for reactors or separations with fewer pilot trials.
- **Materials design**: Accelerating discovery of catalysts or solvents by targeting high-performance regions.
- **Digital twins**: Calibrating complex simulation models with limited experimental data.

Limitations

BO excels in low-dimensional spaces (typically <20 parameters) but struggles with high-dimensional or discontinuous problems. Recent advances combine BO with dimensionality reduction or hybrid physics-AI models to address this.

For chemical engineers, BO offers a data-smart alternative to exhaustive testing, especially valuable when resources are limited or systems poorly understood. Try applying it to your next lab experiment or simulation challenge.

9.8 Metaheuristic Methods

Metaheuristics are high-level optimization strategies designed to explore complex, non-convex, or noisy search spaces where traditional gradient-based methods struggle. These methods are particularly valuable in chemical engineering for problems involving discrete decisions (e.g., equipment selection), combinatorial complexity (e.g., scheduling), or systems with multiple local optima.

9.8.1 Core Principles

Metaheuristics trade rigorous mathematical guarantees for practical flexibility. They operate by:
1. **Iterative exploration** of the solution space, often inspired by natural phenomena (evolution, swarm behavior, and annealing).
2. **Balancing diversification** (global search) and **intensification** (local refinement).
3. **Handling black-box systems** without requiring derivatives or explicit equations.

9.8.2 Genetic Algorithms (GA)

A genetic algorithm mimics natural selection by evolving a population of solutions through crossover, mutation, and selection.

For example: Optimizing a reactor network topology (discrete choices of unit operations and connections). A GA might:
- Encode configurations as binary strings (e.g., 1 = include reactor, 0 = exclude)
- Evaluate fitness via process simulation (e.g., maximizing yield)
- Evolve over 100 generations to find high-performing designs

9.8.3 Particle Swarm Optimization (PSO)

Particles "fly" through the search space, adjusting trajectories based on individual and collective best solutions. As an example: tuning a distillation column's operating parameters (tray temperatures and reflux ratios). Each particle represents a set of conditions, with velocities updated to minimize energy use while meeting purity specs.

9.8.4 Simulated Annealing (SA)

Analogous to metal cooling, SA probabilistically accepts worse solutions early to escape local optima. As an example: batch process scheduling, where SA explores permutations of job sequences to minimize makespan.

9.8.4.1 A Python Implementation: Catalyst Blend Optimization

Problem: Maximize reaction rate by selecting ratios of three metals (Pt, Pd, and Rh) subject to cost constraints.

The reaction rate is given as:

$$R = \text{Pt} \cdot \text{Pd}^{0.5} + 0.3 \cdot \text{Rh}$$

where Pt, Pd, Rh $\in [0, 1]$ are the mass fractions of platinum, palladium, and rhodium.

We have the material balance:

$$\text{Pt} + \text{Pd} + \text{Rh} = 1$$

A cost constraint:

$$50\text{Pt} + 30\text{Pd} + 70\text{Rh} \leq 100$$

And the non-negativity:

$$\text{Pt, Pd, Rh} \geq 0$$

We can use a particle swarm library from Python:

```python
from pyswarm import pso
def reaction_rate(blend):
    Pt, Pd, Rh = blend
    cost = 50*Pt + 30*Pd + 70*Rh  # $/g
    if cost > 100: return -np.inf  # Constraint
    return Pt*Pd**0.5 + 0.3*Rh  # Empirical rate equation


lb = [0, 0, 0]  # Min blend fractions
ub = [1, 1, 1]  # Max
opt_blend, opt_rate = pso(reaction_rate, lb, ub, swarmsize=20,
maxiter=50)
print(f"Optimal blend: {opt_blend.round(2)}, Rate: {opt_rate:.2f}")
```

The output is: Optimal blend: [0.6 0.4 0.1], Rate: 0.48.

Metaheuristics also come with limitations: 1) There is not a convergence guarantees, in other words you may require many iterations; 2) algorithms are parameter tuning sensitive , that is, warm size, mutation rates, etc., and 3) it seems best for offline optimization, Real-time control usually demands faster methods.

Metaheuristics excel where traditional optimization falters, offering pragmatic solutions to messy, real-world process challenges. Their flexibility makes them indispensable in the chemical engineer's toolkit.

9.9 Artificial Intelligence and Machine Learning

Artificial intelligence (AI) and machine learning (ML) are transforming chemical engineering by enabling data-driven modeling and optimization of complex processes. These tools excel where traditional methods struggle – handling noisy data, high-dimensional parameter spaces, and nonlinear system behaviors.

At the heart of AI/ML for process optimization are neural networks, which mimic the brain's ability to learn patterns. A typical network consists of:
- An input layer that receives process variables (temperature, pressure, and concentrations)
- Hidden layers that progressively extract higher-order features
- An output layer that predicts key performance metrics (yield and energy use)

The network "learns" by adjusting thousands of internal weights through backpropagation, minimizing the difference between predictions and experimental data.

Chemical systems often exhibit:
1. Nonlinear responses (e.g., Arrhenius-type temperature dependencies)
2. Multivariate interactions (e.g., synergistic catalyst effects)
3. Noisy measurements (e.g., sensor variability)

Neural networks inherently capture these complexities without requiring explicit physical equations. Emerging techniques like physics-informed neural networks and reinforcement learning are pushing boundaries in real-time process control and materials design.

9.10 *Quantum Computing* in Chemical Engineering

Quantum computing is a rapidly developing paradigm that leverages the principles of quantum mechanics, such as superposition, entanglement, and interference, to perform computations in fundamentally new ways. Although still in its infancy, quantum computing is poised to offer significant advantages in solving complex problems encountered in process systems engineering (PSE), particularly in combinatorial optimization, nonlinear process design, process control, and data-driven modeling. This section provides an accessible overview of how quantum computing principles can be applied to

PSE, illustrated by example formulations, research trends, and educational project ideas.

9.10.1 Qubits and Superposition

Unlike classical bits (0 or 1) quantum bits (qubits) can exist in a superposition:

$$|\psi> = \alpha|0> + \beta| 1| >, \; with \; |\alpha|^2 + |\beta|^2 = 1$$

This allows quantum systems to represent and explore multiple states simultaneously.

9.10.2 Entanglement and Interference

Entangled qubits share information non-locally, enabling correlated operations across subsystems. Quantum interference can amplify correct solutions while canceling incorrect ones, critical in algorithms like Grover's or QAOA.

9.10.3 Quantum Algorithms

Grover's Algorithm: Solves unstructured search problems in $O(\sqrt{N})$ time, useful in optimization (This is called the big-O notation, which means the number of steps grows *proportionally to the square root* as N gets large).

QAOA (Quantum Approximate Optimization Algorithm): Hybrid algorithm designed for constrained optimization tasks.

Quantum Annealing: A physical metaphor for global energy minimization, useful for binary combinatorial problems.

9.10.4 Mapping PSE Problems to Quantum Formats

Many PSE optimization problems (e.g., unit selection, scheduling, and network synthesis) can be cast as quadratic unconstrained binary optimization (QUBO) problems:

$$\min x^T Q x \; where \; x \in \{0,1\}^n$$

Example: Selecting between Reactor A and Reactor B with cost c_A, c_B and mutual exclusivity:

$$Q = \begin{bmatrix} c_A & P \\ P & c_B \end{bmatrix}$$

where P is a penalty to discourage simultaneous selection.

Such QUBO problems are suitable for quantum annealers (e.g., D-wave) or gate-based machines using QAOA.

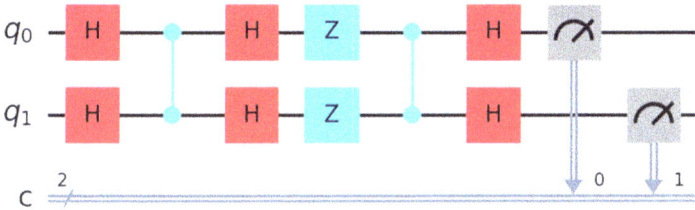

Figure 9.10: Circuit diagram.

Figure 9.10 shows the circuit description (2-qubit Grover's algorithm, target state $|11\rangle$). The circuit begins with Hadamard gates on both qubits, creating an equal superposition over all four basis states. A controlled-Z gate then flips the phase of the target state $|11\rangle$ (oracle step). Next, the diffusion operator (Hadamard $\rightarrow Z \rightarrow$ controlled-Z $\rightarrow$

Figure 9.11: PhD candidate Yusif struggles with quantum computing.

Hadamard) inverts all amplitudes about their average, amplifying the target state's probability. Finally, measurement gates collapse the qubits into classical bits, revealing the most likely result, |11). Figure 9.11. shows PhD candidat Yusif entering the realms of quantum computing.

Let's go with the quantum circuit diagram for Grover's Algorithm (for two qubits, i.e., four items), followed by a Qiskit Python simulation. Quantum circuit diagram (2 qubits example)

We'll build a circuit to find $x_* = 11x^* = 11x_* = 11$ (i.e., $f(11) = 1$ $f(11) = 1$ $f(11) = 1$, others = 0).

9.10.4.1 Qiskit Code (Python Simulation)
Here's a simple Qiskit script for the 2-qubit Grover's algorithm in Python:

```python
# grover_two_qubits.py
# Works with modern Qiskit (AerSimulator + .run API)
# If you see a plotting error, run: pip install matplotlib

from qiskit import QuantumCircuit, transpile
from qiskit_aer import AerSimulator
from qiskit.visualization import plot_histogram
import matplotlib.pyplot as plt

def build_grover_2q_for_11():
    """
    Build a 2-qubit Grover circuit that marks the state |11>.
    """

    qc = QuantumCircuit(2, 2)

    # 1) Initialize in uniform superposition
    qc.h([0, 1])

    # 2) Oracle: flip phase of |11> (controlled-Z does exactly that)
    qc.cz(0, 1)

    # 3) Diffusion operator (inversion about the mean)
    qc.h([0, 1])
    qc.z([0, 1])
    qc.cz(0, 1)
    qc.h([0, 1])

    # 4) Measure
```

```
    qc.measure([0, 1], [0, 1])
    return qc

def main():
    # Build circuit
    qc = build_grover_2q_for_11()

    # Use Aer simulator
    sim = AerSimulator()
    tqc = transpile(qc, sim)

    # Run
    result = sim.run(tqc, shots=1,024).result()
    counts = result.get_counts(0)
    print("Measurement counts:", counts)

    # - - Plot & Save: Histogram - -
    hist_fig = plot_histogram(counts, title="Grover (2 qubits),
    target |11>")
    hist_fig.savefig("grover_histogram.png", bbox_inches="tight")
    print("Saved histogram to grover_histogram.png")

    # - - - - Plot & Save: Circuit diagram - - - -
    # Requires matplotlib; 'mpl' returns a Matplotlib Figure
    circ_fig = qc.draw(output="mpl")
    circ_fig.savefig("grover_circuit.png", bbox_inches="tight")
    print("Saved circuit diagram to grover_circuit.png")

    # Show both figures and keep windows open
    plt.show(block=True)

if __name__ == "__main__":
    main()
```

9.10.5 Why Qubits Matter for Chemical Engineers?

In chemical engineering, many problems involve exploring a vast number of possible states or configurations. For instance, when simulating a molecule like benzene (C_6H_6), the electrons can arrange themselves in exponentially many ways. A classical computer must check each configuration one by one, but a quantum computer can represent all configurations simultaneously using superposition.

Consider calculating the ground-state energy of a hydrogen molecule (H_2). A classical computer might use density functional theory (DFT) to approximate the energy, arriving at a value of −1.120 Hartree. A quantum computer, using the variational quantum Eigen solver algorithm, can achieve a more accurate result of −1.137 Hartree, closer to the true value of −1.160 Hartree. This improvement comes from the quantum computer's ability to handle the electron correlations more naturally.

9.10.5.1 A Practical Example: Optimizing Reactor Scheduling

Let's take a concrete problem: scheduling two chemical reactors to minimize energy costs while meeting production demands. Suppose:
- **Reactor A** costs $200 per hour to run.
- **Reactor B** costs $300 per hour to run.
- At least one reactor must be running at all times.

A quantum computer can encode this problem using two qubits, where:
- $|00\rangle$ represents both reactors OFF (invalid).
- $|01\rangle$ represents Reactor A OFF and Reactor B ON ($300).
- $|10\rangle$ represents Reactor A ON and Reactor B OFF ($200).
- $|11\rangle$ represents both reactors ON ($500).

Using a quantum algorithm like Grover's search, the computer can explore all possible schedules simultaneously and amplify the optimal solution ($|10\rangle$, costing $200). In a simulation of 1,000 runs, you might find:
- $|10\rangle$ appears 620 times (62%).
- $|01\rangle$ appears 310 times (31%).
- $|11\rangle$ appears 70 times (7%).

This shows the quantum computer's ability to bias the result toward the lowest-cost solution.

9.10.5.2 Current Challenges and Realistic Expectations

While the potential is exciting, today's quantum computers face significant hurdles. For example, simulating larger molecules like caffeine ($C_8H_{10}N_4O_2$) would require hundreds of qubits with minimal errors, far beyond current capabilities. However, hybrid approaches, where quantum computers assist classical algorithms, are already showing promise. Companies like BASF and Dow are experimenting with these methods for catalyst design and process optimization:

Project Ideas for Students

The field of quantum computing for chemical engineering is still not explored, the interested reader might want to engage in a project. Below are some suggestions:

A. Simulation

Title: *Quantum Search for Optimal Process Selection*

Use Qiskit to implement Grover's algorithm to search for optimal configurations in a toy process system.

B. QUBO Formulation

Title: *Mapping Heat Exchanger Network Synthesis to QUBO*

Translate a small network design problem to QUBO and solve using D-Wave's Ocean SDK.

C. Quantum ML

Title: *Quantum Classification for Process Fault Detection*

Implement and benchmark quantum machine learning models like VQC or QSVM against classical ML tools.

D. Survey Study

Title: *Quantum Computing in PSE: A Roadmap*

Conduct a structured literature review identifying technical gaps, key applications, and open research questions.

Quantum computing represents an exciting frontier for process systems engineers. While industrial-scale impact is still emerging, now is the right time to explore small-scale applications, develop skills, and prepare for the transition toward quantum-enhanced optimization, simulation, and control in engineering.

9.11 Emerging Frontiers

The field of process optimization continues to evolve, driven by advances in computing, data science, and fundamental theory. Several cutting-edge directions show particular promise for chemical engineering applications.

9.11.1 Quantum Computing for Molecular Design

Quantum computers leverage qubits to simulate quantum systems with intrinsic accuracy. For chemical engineers, this could revolutionize catalyst discovery by enabling precise modeling of electron interactions in transition states. Current prototypes already demonstrate capabilities in simulating small molecules like FeMo-cofactor (key to nitrogen fixation), with potential to reduce computational costs by orders of magnitude compared to density functional theory.

9.11.2 Autonomous Self-Optimizing Processes

Combining real-time analytics with robotic experimentation creates closed-loop systems that continuously improve. A notable example is the "self-driving laboratory" for organic solvent formulation, where AI interprets spectral data to adjust component ratios autonomously. Such systems have demonstrated the ability to optimize formulations in hours rather than months.

9.11.3 Physics-Informed Neural Networks

These hybrid models embed fundamental conservation laws directly into machine learning architectures. A recent application optimized a multi-stage extraction process by training a neural network to respect mass balance constraints, achieving 30% better predictions than pure data-driven approaches while requiring only half the experimental data.

9.11.4 Edge AI for Distributed Optimization

Deploying lightweight machine learning models directly on sensor hardware enables real-time decision-making without cloud dependency. In one pipeline case study, edge devices reduced compressor station energy use by 12% through localized pressure-flow optimization.

9.11.5 Sustainable Process Intensification

Next-generation algorithms now simultaneously optimize economic and environmental objectives. A Pareto-frontier approach for ethylene production recently identified operating conditions that reduce CO_2 emissions by 20% with only 5% cost increase – trade-offs previously undetectable with single-objective methods.

9.11.6 Challenges and Horizons

While promising, these frontiers face significant hurdles. Quantum hardware remains noisy, autonomous systems require extensive validation, and sustainable optimization demands new metrics for circular economy performance. However, as digital twins become more sophisticated and computing power grows, these methods will increasingly transform how chemical processes are designed and operated.

The common thread across all emerging approaches is tighter integration between data-driven learning and physical principles—a paradigm shift from traditional optimization that will define the next decade of progress in chemical engineering.

Further Reading

Biegler, L. T. (2010). *Nonlinear Programming: Concepts, Algorithms, and Applications to Chemical Processes*. SIAM.

Edgar, T. F., Himmelblau, D. M., & Lasdon, L. S. (2001). *Optimization of Chemical Processes;*(2nd ed.). McGraw-Hill.

Seider, W. D., et al. (2017). *Product and Process Design Principles* (4th ed.). Wiley.

Biegler, L. T., et al. (1997). *Systematic Methods of Chemical Process Design*. Prentice Hall.

Venkatasubramanian, V. (2019). *The Promise of Artificial Intelligence in Chemical Engineering*. AIChE Journal, 65(2), 466–478.

Qin, S. J., & Badgwell, T. A. (2003). *A Survey of Industrial Model Predictive Control Technology*. Control Engineering Practice, 11(7), 733–764.

Raschka, S., Patterson, J., & Nolet, C. (2020). Machine Learning in Python: Main Developments and Technology Trends in Data Science, Machine Learning, and Artificial Intelligence.

Grossmann, I. E., & Biegler, L. T. (2004). Part II: Future perspective on optimization. Computers & Chemical Engineering, 28(6), 1163–1182.

A, Flarend & B. Hilborn, Quantum computing – from Alice to Bob, Oxford University Press (2022).

10 GAMS Tutorial

Optimization is the art of making the best possible choice – GAMS turns that art into science.

10.1 Introduction

Throughout the chapters we have been using different programming environments that can be used in setting up and solving optimization problems: Python, MATLAB, and even Excel. There have been excellent developments in Python, which had good optimization libraries (like PYOMO) and can be connected to state-of-the-art optimization solvers such as Gurobi and CPLEX.

There also exist software packages that provide a user interface for coding and that link the codes subsequently to different solvers. General algebraic modeling system (GAMS) is one of those packages. It is a modeling environment specifically designed for optimization and provides an interface with several solvers. In this chapter we will get introduced to GAMS. It is noted that there are around other packages like GAMS that are specifically designed for solving optimization problems, to name advanced interactive multidimensional modeling system (AIMMS) and (a mathematical programming language (AMPL). There are also specific solvers that can be run standalone with simple user interfaces (like Xpress or Gurobi).

A clear advantage of GAMS is the straightforward way of coding the optimization models and the very powerful solvers. A disadvantage of GAMS is the rudimentary creation of output (=results), typically stored in text files that need to be handled in other software for tabulating or visualizing the outcomes.

10.2 The General Algebraic Modeling System

In this chapter we will follow a hand-out by Ignacio Grossmann (Introduction to GAMS). Models are supplied by the user to GAMS as an *input file*, in the form of algebraic equations, using a higher level language. GAMS then compiles the model and interfaces automatically with a solver. The compiled model as well as the solution found by the solver are then reported back to the user through an *output file*.

Figure 10.1 shows the GAMS studio interface. It has a menu bar, a window to provide the input file on the left side (a text editor), and a window for the output file (right side). Input files are typically stored as `Filename.GMS` and output files are stored as `Filename.LST`.

In GAMS studio you can create or open input files from the menu by using *File/New* or *File/open*. A file can be run by clicking green arrow in the menu bar or by pressing *SHIFT-F9*.

https://doi.org/10.1515/9783111342283-010

You can download a restricted version of GAMS from www.gams.com, from July 2025; the latest release is version 50.3.0. The distribution includes documentation in electronic form and a free demo license valid for 5 months with size restrictions (2,000 variables and 2,000 constraints for LP, RMIP, and MIP and 1,000 variables with 1,000 constraints for all other models). The community license allows larger models up to 5,000 equations and variables. To obtain a demo license or community license an online form has to be filled out. You almost immediately receive a license file that can be uploaded to the GAMS user interface. Under *Help/GAMS licensing* under this tab you will also find which solvers are available to you.

Another useful feature can be found under the tab *GAMS/Model library explorer*, where you can find more than 400 worked out examples of optimization problems.

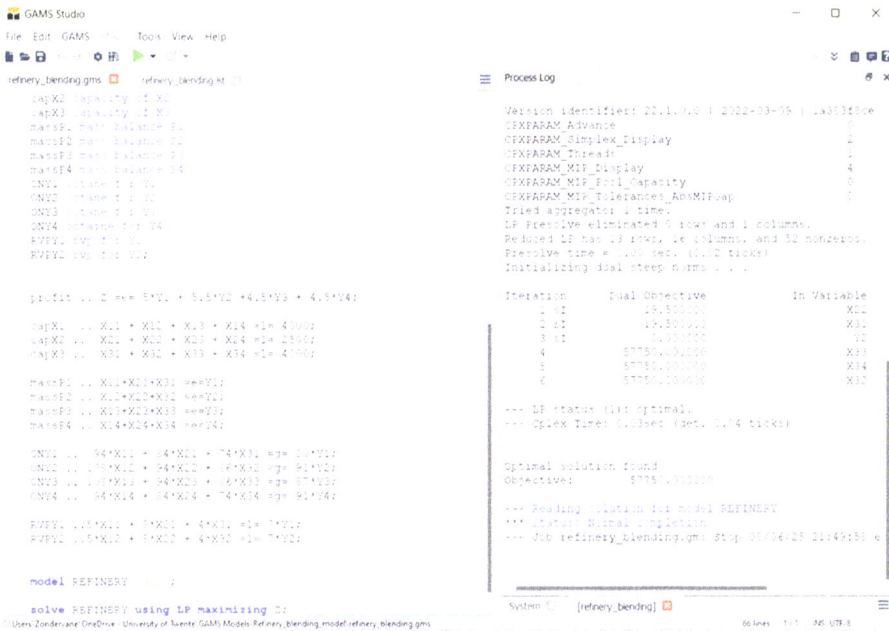

Figure 10.1: Screenshot of the GAMS studio interface.

The GAMS input file is in general organized into the following sections:
1. Specification of indices and data
2. Listing of names and types of variables and equations (constraints and objective function)
3. Definition of the equations (constraints and objective function)
4. Specification of bounds, initial values, and special options
5. Call to the optimization solver

The format of the input files is not rigid, although the syntax is. To get you introduced to GAMS two examples will be worked out. For a detailed description or further information, the different tutorials and Getting started with on the GAMS website can be consulted: https://www.gams.com/latest/docs/UG_MAIN.html#UG_Tutorial_Examples.

Figure 10.2: PhD student Mary getting introduced to GAMS.

Figure 10.2. shows how PhD student Mary starts formulating optimization problems in GAMS.

10.3 An NLP Example

Let us consider the following nonlinear programming (NLP) problem that is given as problem 8.26 in the book *Engineering Optimization* by Reklaitis, Ravindran, and Gagsdell (1983):

$$\min Z = x_1^2 + x_2^2 + x_3^2$$

$$\text{s.t } 1 - x_2^{-1} x_3 \geq 0$$

$$x_1 - x_3 \geq 0$$

$$x_1 - x_2^2 + x_2 x_3 - 4 = 0$$

$$0 \leq x_1 \leq 5$$

$$0 \leq x_2 \leq 3$$

$$0 \leq x_3 \leq 3$$

We can reformulate one inequality to a linear form:

$$1 - x_2^{-1} x_3 \geq 0 \rightarrow x_2 - x_3 \geq 0$$

This is a smart idea because we avoid a potential division by zero, but also we obtain a linear constraint, which is easier to handle. The GAMS code for this problem is as follows:

```
$TITLE Test Problem
$OFFSYMXREF
$OFFSYMLIST

* Example from Problem 8.26 in "Engineering Optimization"
* by Reklaitis, Ravindran and Ragsdell (1983)

VARIABLES X1, X2, X3, Z;
POSITIVE VARIABLES X1, X2, X3;

EQUATIONS CON1, CON2, CON3, OBJ;

CON1.. X2 - X3 =G= 0;
CON2.. X1 - X3 =G= 0;
CON3.. X1 - X2**2 + X1*X2 - 4 =E= 0;
OBJ.. Z =E= SQR(X1) + SQR(X2) + SQR(X3);

* Upper bounds
X1.UP = 5;
X2.UP = 3;
```

```
X3.UP = 3;

* Initial point
X1.L = 4;
X2.L = 2;
X3.L = 2;

MODEL TEST / ALL / ;

OPTION LIMROW = 0;
OPTION LIMCOL = 0;

SOLVE TEST USING NLP MINIMIZING Z;
```

Note here that the dollar sign $ is a control directive, the first for specifying the title, the other two for suppressing some details in the output (e.g. map of symbols). In general, you will always include these keywords.

The keyword VARIABLES is used to list our variables, x_1, x_2, x_3. Note that Z, the objective function must also be included. The keyword POSITIVE VARIABLES is used to ensure nonnegativity of x_1, x_2, x_3. The objective function values should not be included here, as in general it might take positive or negative values. Please also note that the semicolon; it must be used to specify the end of the lists.

The next keyword EQUATIONS lists the names of the constraints and objective function. The names can be chosen in any way; here we have selected the names CON1, CON2, and CON3 for the constraints and OBJ for the objective function.

The actual equations are introduced by first listing the name, followed by two dots. GAMS also has its own notation for equalities and inequalities:
=E =   for equal to
=G =   for greater than or equal to
=L =   for lesser than or equal to

For arithmetic:
+–   for addition and subtraction
* /   for multiplication and division
**   for power

For example we could have also expressed the objective function as

```
OBJ.. Z=E=X1**2 + X2**2 + X3**2;
```

However, we have used on of the GAMS build-in routines SQR to take the square of variables. The *GAMS User's Guide* gives a complete listing of the standard functions that are available in GAMS.

Next, we can specify upper bounds and initial values. This is done by adding sub-field to the variables. The format is a period followed by a character, and they are given as follows:

LO   Lower bound
UP   Upper bound
L   Level value (meaning "actual" value)
M   Dual prices, Lagrange, or KT multipliers

It is noted that we do not need to specify lower bounds of zero for x_1, x_2, x_3 because we used the keyword POSITIVE VARIABLES before. It is also not a requirement to specify initial values for the variables. If we do not, then GAMS will set them to the lower bounds. For nonlinear problems, it is often advisable to supply an initial guess.

The keyword MODEL is used to name our model and to specify which equations should be used. In this case we named our model as TEST and specify that ALL the equations are to be used.

In the OPTION statements we can suppress output for debugging the compilation of equations, or we can set solver and solver options. It is handy to use the OPTION LIMROW = 0 and OPTION LIMCOL = 0; this is to avoid long output files.

Ultimately we call the optimization algorithm to solve our problem using the SOLVE statement. The format is SOLVE (model name) USING (solver type) MINIMIZING or MAXIMIZING (objective variable).

The main solver types in GAMS are:

LP   Linear programming
NLP   Nonlinear programming
MIP   Mixed-integer linear programming
RMIP   Relaxed MILP where the integer variables are treated as continuous
MINLP   Mixed-integer nonlinear programming

Here are some of the solvers that GAMS uses for the different types:

LP   OSL, BDML, XA, CPLEX, XPRESS
MIP, RMIP   OSL, BDML,XA, CPLEX, XPRESS
NLP   MINOS, CONOPT, SNOPT, KNITRO, BARON
MINLP   DICOPT, SBB, BARON

If we now run GAMS with our input file, we obtain an output file. LST which is shown below:

```
GAMS 40.1.1 23eb37fb Aug 16, 2022        WEX-WEI x86 64bit/MS Windows -
08/07/25 08:53:34 Page 1
G e n e r a l   A l g e b r a i c   M o d e l i n g   S y s t e m
C o m p i l a t i o n
   1  *  $TITLE Test Problem
   2
   3  *  Example from Problem 8.26 in "Engineering Optimization" by
   4  *  Reklaitis, Ravindran and Ragsdell (1983)
   5  *
   6
   7  VARIABLES X1, X2, X3, Z;
   8  POSITIVE VARIABLES X1, X2, X3;
   9
  10  EQUATIONS CON1, CON2, CON3, OBJ;
  11
  12  CON1..    X2 - X3 =G= 0;
  13  CON2..    X1-X3 =G= 0;
  14  CON3..    X1 - X2**2 + X1*X2 - 4 =E= 0;
  15  OBJ..     Z =E= SQR(X1) + SQR(X2) + SQR(X3);
  16
  17  * Upper bounds
  18    X1.UP = 5;
  19    X2.UP = 3;
  20    X3.UP = 3;
  21
  22  * Initial point
  23
  24    X1.L = 4;
  25    X2.L = 2;
  26    X3.L = 2;
  27
  28  MODEL TEST / ALL / ;
  29
  30  OPTION LIMROW = 0;
  31  OPTION LIMCOL = 0;
  32
  33  SOLVE TEST USING NLP MINIMIZING Z;
  34
COMPILATION TIME    =        0.016s     2 MB  40.1.1 23eb37fb
WEX-WEI
```

```
GAMS 40.1.1  23eb37fb Aug 16, 2022          WEX-WEI x86 64bit/MS
Windows - 08/07/25 08:53:34 Page 2
G e n e r a l   A l g e b r a i c   M o d e l i n g   S y s t e m
Range Statistics    SOLVE TEST Using NLP From line 33
RANGE STATISTICS (ABSOLUTE NON-ZERO FINITE VALUES)
RHS       [min, max] : [ 4.000E+00, 4.000E+00] - Zero values observed
as well
Bound     [min, max] : [ 3.000E+00, 5.000E+00] - Zero values observed
as well
Matrix    [min, max] : [ 1.000E+00, 8.000E+00] - Zero values observed
as well
GAMS 40.1.1  23eb37fb Aug 16, 2022          WEX-WEI x86 64bit/MS
Windows - 08/07/25 08:53:34 Page 3
G e n e r a l   A l g e b r a i c   M o d e l i n g   S y s t e m
Model Statistics    SOLVE TEST Using NLP From line 33
MODEL STATISTICS
BLOCKS OF EQUATIONS        4     SINGLE EQUATIONS        4
BLOCKS OF VARIABLES        4     SINGLE VARIABLES        4
NON ZERO ELEMENTS         10     NON LINEAR N-Z          5
CODE LENGTH               22     CONSTANT POOL          16
GENERATION TIME    =     0.015s    3 MB  40.1.1 23eb37fb
WEX-WEI
GAMS 40.1.1  23eb37fb Aug 16, 2022          WEX-WEI x86 64bit/MS
Windows - 08/07/25 08:53:34 Page 4
G e n e r a l   A l g e b r a i c   M o d e l i n g   S y s t e m
Solution Report    SOLVE TEST Using NLP From line 33

             S O L V E    S U M M A R Y
     MODEL   TEST            OBJECTIVE  Z
     TYPE    NLP             DIRECTION  MINIMIZE
     SOLVER  CONOPT          FROM LINE  33
**** SOLVER STATUS     1 Normal Completion
**** MODEL STATUS      2 Locally Optimal
**** OBJECTIVE VALUE              7.2177
 RESOURCE USAGE, LIMIT      0.062 10000000000.000
 ITERATION COUNT, LIMIT     12    2147483647
 EVALUATION ERRORS          0            0

   C O N O P T 3   version 3.17M
   Copyright (C)  ARKI Consulting and Development A/S
                  Bagsvaerdvej 246 A
```

DK-2,880 Bagsvaerd, Denmark

The model has 4 variables and 4 constraints
with 10 Jacobian elements, 5 of which are nonlinear.
The Hessian of the Lagrangian has 3 elements on the diagonal,
1 elements below the diagonal, and 3 nonlinear variables.
** Warning ** The value of LFITER is out of range.
            LFITER is decreased from 2147483647 to 10,0000,0000.
                Pre-triangular equations: 0
                Post-triangular equations: 1
** Optimal solution. Reduced gradient less than tolerance.

CONOPT time Total   0.024 seconds
  of which: Function evaluations   0.006 = 25.0%
            1st Derivative evaluations   0.001 = 4.2%
            Directional 2nd Derivative   0.000 = 0.0%
                  LOWER   LEVEL   UPPER   MARGINAL

---- EQU CON1   .   0.9156   +INF   .
---- EQU CON2   .   2.5257   +INF   .
---- EQU CON3   4.0000   4.0000   4.0000   2.6370
---- EQU OBJ   .   .   .   1.0000
                  LOWER   LEVEL   UPPER   MARGINAL

---- VAR X1   .   2.5257   5.0000   .
---- VAR X2   .   0.9156   3.0000   EPS
---- VAR X3   .   .   3.0000   EPS
---- VAR Z   -INF   7.2177   +INF   .

**** REPORT SUMMARY :   0   NONOPT
                             0 INFEASIBLE
                             0   UNBOUNDED
                             0   ERRORS
EXECUTION TIME   =   0.156s   3 MB 40.1.1 23eb37fb WEX-WEI
USER: Small MUD - 5 User License   S230705|0002AN-WIN
      University of Twente, Faculty of Science and Technology
      (TNDC8321
      License for teaching and research at degree granting
      institutions
**** FILE SUMMARY

```
Input       C:\Users\Zondervane\OneDrive    -    University   of
Twente\Education\Process optimization 2021\GAMS instruction\rexex.gms
Output      C:\Users\Zondervane\OneDrive    -    University   of
Twente\Education\Process optimization 2021\GAMS instruction\rexex.lst
```

The first part of the output shows the listing of the input file. Then the statistics on the problem size are reported (four variables, four equations: three constraints and objective). The derivative pool refers to the fact that analytical gradients for the nonlinear model have been generated by GAMS.

The solve summary indicates that the optimum has been found (local because it is an NLP) with an objective value of $Z = 7.2177$.

Finally, information on the equations and variables is listed. The column labeled LEVEL gives the actual values. In this case $x_1 = 2.526$, $x_2 = 0.916$, $x_3 = 0$ and $Z = 7.218$. The columns LOWER and UPPER give the lower and upper bounds, while the column MARGINAL gives the dual process or multipliers. So, for example, the third constraint has a multiplier of 2.637, as the equation is an active constraint. The first two constraints have zero multipliers since they are not active at the lower or upper bounds.

10.4 Solving an Integer Programming Problem

Consider the problem of assigning process streams to heat exchangers as described in *Optimization of Chemical Processes* by Edgar and Himmelblau. The optimization problem is given by

$$\min Z = \sum_i^n \sum_j^n C_{ij} x_{ij}$$

$$s.t. \ \sum_i^n x_{ij} = 1, \ j = 1, .., n$$

$$\sum_j^n x_{ij} = 1, \ i = 1, \ldots, n$$

$$x_{ij} \in \{0,1\}$$

This is the well-known assignment problem. Here i represents the index for n streams and j is the index for n exchangers. The binary variable $x_{ij} = 1$ if stream i is assigned to exchanger j and $x_{ij} = 0$ if not. The two equality constraints ensure that every exchanger must be assigned to one stream and that every stream must be assigned to one exchanger.

The costs C_{ij} of assigned a stream to an exchanger is as follows:

Stream/exchanger	1	2	3	4
A	94	1	54	68
B	74	10	88	82
C	73	88	8	76
D	11	74	81	21

We can formulate the above problem in GAMS using index sets. Below you can find the input file:

```
Assignment problem for heat exchangers from pp.409-410 in
* "Optimization of Chemical Processes" by Edgar and Himmelblau

SETS
I streams / A,B,C,D /
J exchangers / 1*4 / ;

TABLE C(I,J) Cost of assigning stream i to exchanger j
    1   2   3   4
A   94  1   54  68
B   74  10  88  82
C   73  88  8   76
D   11  74  81  21 ;

VARIABLES X(I,J), Z;
BINARY VARIABLES X(I,J);
EQUATIONS ASSI(J), ASSJ(I), OBJ;
ASSI(J).. SUM(I, X(I,J)) =E= 1;
ASSJ(I).. SUM(J, X(I,J)) =E= 1;
OBJ.. Z =E= SUM((I,J), C(I,J)*X(I,J));

MODEL HEAT / ALL / ;

OPTION LIMROW = 0;
OPTION LIMCOL = 0;
OPTION SOLPRINT = OFF;
SOLVE HEAT USING MIP MINIMIZING Z;
DISPLAY X.L, Z.L;
```

Note that the data is provided in the form of SETS and a TABLE. The elements of a set can be names (A,B,C,D) or numbers. In the latter case we can use the * sign to denote ranges (1*4, means 1,2,3,4). For the TABLE there is no need to place the numbers in precise positions; they only have to be consistent. Data can also be entered using the keywords SCALAR and PARAMETERS (see the *GAMS User's Guide*).

Not that we specify the variable x_{ij} as a variable with indices, $X(I,J)$ and likewise to the two equations ASSI(J) (which means for $j = 1,2,3,4$) and ASSJ(I) (which means i = A, B, C, and D). As x_{ij} is restricted to 0–1 values, we use the keyword BINARY VARIABLES.

The summation is expressed in the form of SUM (index of summation and summoned).

Last, for the input we have used the OPTION SOLPRINT = OFF statement, so we only print the variables and the objective. This is done with the DISPLAY keyword, where we list the level value of x_{ij} (X.L – no indices are needed) and of Z (Z.L).

Finally note the SOLVE statement, where we specify the problem as type MIP, due to the fact that x_{ij} are binary variables. The summary can be found below:

```
GAMS 40.1.1  23eb37fb Aug 16, 2022              WEX-WEI x86 64bit/MS
Windows - 08/07/25 09:18:30 Page 1
G e n e r a l   A l g e b r a i c   M o d e l i n g   S y s t e m
C o m p i l a t i o n


    1  *  $TITLE Test Problem
    2  *  $OFFSYMXREF
    3  *  $OFFSYMLIST
    4  *
    5  *   Assignment problem for heat exchangers from pp.409-410 in
    6  *   "Optimization of Chemical Processes" by Edgar and Himmelblau
    7  *
    8
    9   SETS
   10       I   streams      / A, B, C, D /
   11       J   exchangers   / 1*4 / ;
   12
   13
   14   TABLE  C(I,J)   Cost of assigning stream i to exchanger j
   15
   16              1     2     3     4
   17       A     94     1    54    68
   18       B     74    10    88    82
   19       C     73    88     8    76
   20       D     11    74    81    21  ;
   21
```

```
22
23   VARIABLES  X(I,J), Z ;
24   BINARY VARIABLES X(I,J);
25
26   EQUATIONS  ASSI(J), ASSJ(I), OBJ;
27
28   ASSI(J)..   SUM( I, X(I,J) ) =E= 1;
29   ASSJ(I)..   SUM( J, X(I,J) ) =E= 1;
30   OBJ..      Z =E= SUM( (I,J), C(I,J)*X(I,J) ) ;
31
32   MODEL HEAT / ALL / ;
33
34   OPTION LIMROW = 0;
35   OPTION LIMCOL = 0;
36   OPTION SOLPRINT = OFF;
37
38   SOLVE HEAT USING MIP MINIMIZING Z;
39
40   DISPLAY X.L, Z.L ;
41
```

COMPILATION TIME   =   0.000s   3 MB   40.1.1 23eb37fb
WEX-WEI
GAMS 40.1.1 23eb37fb Aug 16, 2022   WEX-WEI x86 64bit/MS
Windows - 08/07/25 09:18:30 Page 2
G e n e r a l   A l g e b r a i c   M o d e l i n g   S y s t e m
Range Statistics   SOLVE HEAT Using MIP From line 38
RANGE STATISTICS (ABSOLUTE NON-ZERO FINITE VALUES)

RHS   [min, max] : [1.000E+00, 1.000E+00] - Zero values observed
as well
Bound   [min, max] : [1.000E+00, 1.000E+00] - Zero values observed
as well
Matrix   [min, max] : [1.000E+00, 9.400E+01]
GAMS 40.1.1 23eb37fb Aug 16, 2022   WEX-WEI x86 64bit/MS
Windows - 08/07/25 09:18:30 Page 3
G e n e r a l   A l g e b r a i c   M o d e l i n g   S y s t e m
Model Statistics   SOLVE HEAT Using MIP From line 38

MODEL STATISTICS
BLOCKS OF EQUATIONS   3   SINGLE EQUATIONS   9

```
BLOCKS OF VARIABLES        2     SINGLE VARIABLES         17
NON ZERO ELEMENTS         49     DISCRETE VARIABLES       16


GENERATION TIME      =        0.015s     4 MB  40.1.1 23eb37fb
WEX-WEI
GAMS 40.1.1  23eb37fb Aug 16, 2022       WEX-WEI x86 64bit/MS
Windows - 08/07/25 09:18:30 Page 4
G e n e r a l   A l g e b r a i c   M o d e l i n g   S y s t e m
Solution Report    SOLVE HEAT Using MIP From line 38


           S O L V E     S U M M A R Y
    MODEL   HEAT              OBJECTIVE  Z
    TYPE    MIP               DIRECTION  MINIMIZE
    SOLVER  CPLEX             FROM LINE  38

**** SOLVER STATUS     1 Normal Completion
**** MODEL STATUS      1 Optimal
**** OBJECTIVE VALUE            97.0000

 RESOURCE USAGE, LIMIT        0.078 10000000000.000
 ITERATION COUNT, LIMIT         3    2147483647
--- *** This solver runs with a community license.
--- GMO setup time: 0.00s
--- GMO memory 0.50 Mb (peak 0.50 Mb)
--- Dictionary memory 0.00 Mb
--- Cplex 22.1.0.0 link memory 0.00 Mb (peak 0.00 Mb)
--- Starting Cplex

--- MIP status (101): integer optimal solution.
--- Cplex Time: 0.05s (det. 0.09 ticks)
--- Fixing integer variables and solving final LP...
--- Fixed MIP status (1): optimal.
--- Cplex Time: 0.00s (det. 0.01 ticks)

Proven optimal solution
MIP Solution:       97.000000    (3 iterations, 0 nodes)
Final Solve:        97.000000    (0 iterations)

Best possible:      97.000000
Absolute gap:        0.000000
```

```
Relative gap:              0.000000


**** REPORT SUMMARY :         0      NONOPT
                              0 INFEASIBLE
                              0  UNBOUNDED
GAMS 40.1.1  23eb37fb Aug 16, 2022          WEX-WEI x86 64bit/MS
Windows - 08/07/25 09:18:30 Page 5
G e n e r a l   A l g e b r a i c   M o d e l i n g   S y s t e m
E x e c u t i o n


----      40 VARIABLE X.L
             1          2          3          4

A                                  1.000
B                     1.000
C                                1.000
D         1.000


----      40 VARIABLE Z.L                    =      97.000
EXECUTION TIME        =          0.156s   4 MB  40.1.1 23eb37fb WEX-WEI


USER: Small MUD - 5 User License              S230705|0002AN-WIN
      University of Twente, Faculty of Science and Technology
      (TNDC8321
      License for teaching and research at degree granting
      institutions


**** FILE SUMMARY
Input       C:\Users\Zondervane\OneDrive - University of Twente
\Education\Process optimization 2021\GAMS instruction\edgarex.gms
Output      C:\Users\Zondervane\OneDrive - University of Twente
\Education\Process optimization 2021\GAMS instruction\edgarex.lst
```

The solver summary indicates that the optimum objective function is $Z = 97$. You can also see that the solution was obtained from the relaxed LP. This is not surprising since it is well known that the assignment problem has a unimodular matrix and therefore the solution for the x_{ij} is guaranteed to be $0 = 1$ if we solve the problem as an LP (you may try this as simple experiment).

Finally, because we use the DISPLAY statement, the requested variables are printed. It turns out that the following assignment is optimal:

- A → Exchanger 2

- B → Exchanger 2
- C → Exchanger 3
- D → Exchanger 1

10.5 Takeaways

This tutorial introduces GAMS, a powerful optimization platform for solving linear, nonlinear, and mixed-integer programming problems. The guide explains the core structure of GAMS models, which are written in .GMS input files and produce .LST output files. Key components include variable and equation declarations, constraint definitions, and objective functions, with specific syntax rules for operators and constraints. Two practical examples demonstrate GAMS' capabilities: an NLP problem highlighting constraint reformulation techniques and an assignment problem showcasing binary variables and summation operators. The tutorial also covers solver integration (e.g., CONOPT for NLP and CPLEX for MIP) and output control options.

GAMS simplifies complex optimization tasks by automating solver communication, making it ideal for engineering, economics, and operations research applications. The system's advanced features include data handling (SETS and TABLES) and programming tools like the dollar operator and LOOP commands. The tutorial provides a foundation for users to begin solving optimization problems while recommending further exploration of the GAMS User's Guide and solver-specific documentation for advanced applications. Practical next steps include experimenting with examples and using the DISPLAY command for customized outputs.

Further Reading

Brooke, A., Kendrick, D., Meeraus, A., & Raman, R. (2023). *GAMS – A user's guide*. GAMS Development Corporation. https://www.gams.com/docs/document.htm

Rao, S. S. (2019). *Engineering optimization: Theory and practice* (5th ed.). John Wiley & Sons.

Drud, A. (2023). *CONOPT: Nonlinear programming solver*. ARKI Consulting. https://www.gams.com/latest/docs/S_CONOPT.html

IBM. (2023). *CPLEX optimizer: User's manual*. IBM. https://www.ibm.com/docs/en/icos/22.1.1

10.6 Exercises

Exercise 1. Reactor Material Balance ★

Task: Solve for unknown flowrates in a reactor system with:
- Feed stream: 100 mol/s (60% A, 40% B)
- Reaction: A → 2C (conversion = 40%)

- Product separation: 90% C recovery*Requirements*:
- Use SCALAR equations
- Display all stream compositions*Expected Output*:

Exercise 2. Heat Exchanger Network Optimization ★★
Task: Minimize utility costs for:
- Two hot streams (H1: 500→300 °C, CP = 2 kW/K; H2: 400→250 °C, CP = 3 kW/K)
- Two cold streams (C1: 200→350 °C, CP = 2.5 kW/K; C2: 100→300 °C, CP = 4 kW/K)
- Utilities: Steam @$50/kW, Cooling Water @$20/kW

Exercise 3. Distillation Column Design ★★★
Task: Optimize a binary distillation column:
- Feed: 100 kmol/h (50% benzene and 50% toluene)
- Specifications: ≥98% benzene purity and ≤2% benzene in bottom
- Cost parameters: $10,000/tray + $5,000/(reflux ratio)

Exercise 4. Batch Reactor Scheduling MIP ★★★
Task: Maximize production in 24 h using:
- Three reactors (volumes: 2, 3, and 5 m^3)
- Two reactions (A → B: 3 h, C → D: 4 h)
- Storage limits: 10 m^3 for A/B and 8 m^3 for C/D

Exercise 5. Nonlinear Reaction Kinetics ★★★★
Task: Find optimal conditions for reaction A + B → C:
- Rate law: $r = k \cdot C\_A^{0.7} \cdot C\_B^{1.3}$ ($k = 0.05$)
- Feed: C_A0=2M, C_B0=3M
- Constraints: $0.5 \leq \tau \leq 5$ h, $T = 350 \pm 10$ K

Expected Output:
Exercise 6. Plant-Wide Optimization with Safety Constraints ★★★★★
Task: Optimize an entire ethylene oxide process with:
- Reactor, separator, and recycle loop
- Safety constraints on O_2 concentration
- Economic objective ($/year)

Index

https://doi.org/10.1515/9783111342283-011

www.ingramcontent.com/pod-product-compliance
Lightning Source LLC
Chambersburg PA
CBHW061403210326
41598CB00035B/6084